The Cambridge Manuals of Science and
Literature

BEES AND WASPS

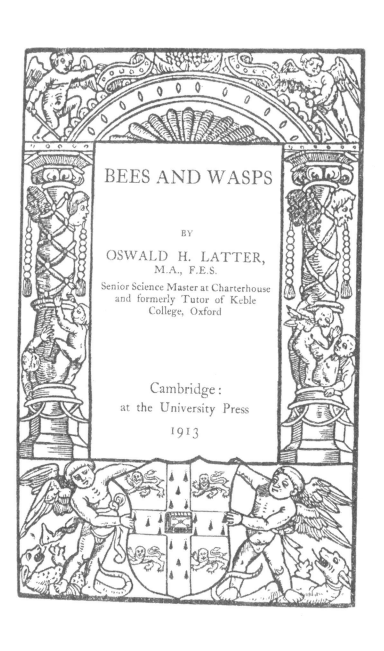

BEES AND WASPS

BY

OSWALD H. LATTER,
M.A., F.E.S.

Senior Science Master at Charterhouse
and formerly Tutor of Keble
College, Oxford

Cambridge:
at the University Press
1913

CAMBRIDGE UNIVERSITY PRESS
Cambridge, New York, Melbourne, Madrid, Cape Town,
Singapore, São Paulo, Delhi, Tokyo, Mexico City

Cambridge University Press
The Edinburgh Building, Cambridge CB2 8RU, UK

Published in the United States of America by Cambridge University Press, New York

www.cambridge.org
Information on this title: www.cambridge.org/9781107605749

© Cambridge University Press 1913

First published 1913
First paperback edition 2011

A catalogue record for this publication is available from the British library

ISBN 978-1-107-60574-9 Paperback

Cambridge University Press has no responsibility for the persistence or
accuracy of URLs for external or third-party internet websites referred to in
this publication, and does not guarantee that any content on such websites is,
or will remain, accurate or appropriate.

*With the exception of the coat of arms
at the foot, the design on the title page is a
reproduction of one used by the earliest known
Cambridge printer, John Siberch,* **1521**

PREFACE

THIS little book makes no claim to be an exhaustive treatment of a very large subject. It is merely a sketch in which I have filled in details of those few insects which happen to come under my immediate notice, have touched lightly on others in mere outline, and have omitted entirely many extensive groups which are visible enough to those who look for them, but which for various reasons are not familiar to the general public. It has been my aim to write mainly on those Hymenoptera which have especially interested and delighted me, in the hope that I might thus best interest others and introduce them to one of the most fascinating branches of Natural History. It is for this reason that I have confined my remarks to British species of Hymenoptera. Much of my information has been derived from other sources: these are not acknowledged individually in the text; but it is hoped that the bibliography at the end of the book will be recognised as an adequate acknowledgement of my

vi   PREFACE

indebtedness. I cannot refrain from taking this opportunity of expressing my gratitude to my old friend and schoolmaster, the late Mr W. H. Poole of Charterhouse, who first introduced me as a boy to this group of insects whose study has since afforded me so much pleasure and recreation both in the laboratory and on the sunny heaths of Surrey and elsewhere. To my friend and pupil Mr Kenneth W. Merrylees I am much indebted for kind help with many of the illustrations.

O. H. L.

CHARTERHOUSE,
    GODALMING.
        February, 1913.

CONTENTS

CHAP.		PAGE
I.	Introduction	1
II.	Fossores or Digger-Wasps: Pompilid section	5
III.	Fossores or Digger-Wasps: Sphegid section	19
IV.	Diploptera (double-winged wasps)	36
V.	Bees (Anthophila, or flower-lovers)	57
VI.	Long-pointed-tongued bees (Apidae)	73
VII.	The social-bees	87
VIII.	Some structural features: the sting and the "tongue"	108
IX.	Collecting and preserving aculeate Hymenoptera	118
	Bibliography	129
	Index	130

LIST OF ILLUSTRATIONS

FIG.		PAGE
1.	*Pompilus viaticus*	14
2.	*Ammophila sabulosa*	21
3.	*Oxybelus uniglumis*	25
4.	Wings of a Hymenopteron and *Crabro*	29
5.	*Odynerus parietum*	37
6.	Nest of *Odynerus*	39
7.	Diagram of wasp's nest at an early stage	43
8.	Diagram of wasp's nest towards the end of summer	46
9.	Back view of mouth parts of *Colletes* and *Andrena*	59
10.	Front view of mouth parts of *Bombus*	61
11.	*Colletes succincta*	62
12.	Last segment of abdomen of female *Halictus*	67
13.	*Dasypoda hirtipes*	71
14.	Portion of a rose-leaf which has been attacked by a leaf-cutter bee (*Megachile*)	79
15.	Three cells of leaf-cutter bee (*Megachile*) in a piece of rotten wood	80
16.	*Eucera longicornis*	86
17.	Honey-bee. A, outer aspect of hind leg. B, inner aspect of metatarsus and lower part of tibia of hind leg. C, junction of tibia and metatarsus of middle leg, showing spur	105
18.	Ventro-lateral view of sting of *Vespa germanica* ♀	109
19.	Diagrammatic cross-section of a wasp's sting	110
20.	Optical section of tip of a needle from the sting of *Vespa germanica* ♀	110
21.	Diagram of side view of bee's head to show disposition of the mouth parts	115

CHAPTER I

INTRODUCTION

THE terms "bee" and "wasp" probably do not, except to an entomologist, denote more than a very small assemblage of insects. The honey-bee and perhaps one or two species of bumble-bees on the one hand, and the yellow and black banded insects that cause such consternation at the breakfast table on the other, represent for most of us the sum total of personal acquaintance in this class of animals. It is therefore necessary at the outset to explain the sense in which we are here employing the two words which constitute the title of this little volume.

There is one feature, and that fortunately and not unnaturally impressed firmly on the popular mind, which is possessed by no other insects except the bees, wasps and their near relations the ants, and by which they may therefore at once be distinguished. This feature is the possession by the female of a sting furnished with a poison bag. Identification by means of this test is however not entirely satisfactory: in the first place the males cannot sting;

secondly, the test is difficult of application to the dead insect; thirdly, most of these stinging ("aculeate") insects are unable to pierce the human skin with their feeble weapon and so fail to give a conclusive answer to the question asked of them; while, fourthly, those provided with a sting sufficiently powerful to gain entrance to our own dermis produce results so unpleasant as to deter any but the most enthusiastic devotee from further enquiry in the same direction. Our readers will therefore pardon us, if, in order to save unnecessary pain and disappointment, we now proceed to enumerate a few of the more conspicuous structural characters by which collectively these interesting insects may be recognised and distinguished from others.

The members of the great Order Hymenoptera (membrane-winged), in which are also included the saw-flies, gall-flies and ichneumon-flies, possess as a rule four membranous and usually transparent wings which are destitute of scales (contrast the Lepidoptera, butterflies and moths), and are of but moderate size, the anterior pair being larger than the posterior (contrast the beetles, grasshoppers, earwigs, etc.). The areas, or "cells," into which the wings are marked out by the nervures or veins are not regular in size and shape, nor do they ever exceed twenty in the front or fifteen in the hind wing (contrast dragon-flies, may-flies, etc.). The disposition of the second

1] INTRODUCTION 3

and third pairs of jaws (first and second maxillae) varies considerably and will be dealt with more fully later; but in all cases the first pair takes the form of well-developed mandibles adapted for biting. In that portion of the stinging (aculeate) section of the Order, with which we are here concerned, the antennae (feelers) of the males have thirteeen joints, while those of the females, whether "queens" or "workers," have but twelve. The ants are readily distinguishable from the other "stingers" by the presence of one or more irregular elevations ("nodes") on the upper surface of the "stalk" or "waist" by which the hindmost portion (abdomen) of the body is united to the middle region or "thorax." But it is no easy matter to distinguish between a "bee" and a "wasp" in the wide sense in which we are now using these words. Structurally, two diagnostic characters may be relied on, so far at any rate as British species are concerned: one lies in the shape of the hairs with which more or less of the body is clad; in "bees" some at least of the hairs are "plumose," i.e. provided with short lateral offsets like those on a sparsely fluffy feather of a bird; whereas in "wasps" all the hairs are "simple," i.e. destitute of offsets. The other is the widened condition of the metatarsal joint (see fig. 17) of the hind legs. Since a microscope of fairly high power is required to render the shape of the hairs visible, and since in some male bees (*Andrena*) the enlargement of the metatarsus is indistinct, we are constrained to fall back on the habits of the insects in an endeavour to discriminate between them. The technical name for the "bees" is *Anthophila* (flower-lovers), and though many of the "wasps" frequent flowers and nourish themselves upon the nectar secreted by them, yet it is the fact that none but the bees provision their nests with the pollen of flowers, honey, etc. for the benefit of their offspring. All the "wasps," notwithstanding that when adult they feed upon plant products, supply their young with animal food, such as spiders, caterpillars, or the flesh of larger carcases. Here then we find a sure, though granted not an easy means of distinguishing between "bee" and "wasp": —the bee grub is nourished upon vegetable products collected by its mother or some other bee, the wasp grub is carnivorous.

The term "wasp" as here employed includes many species of insects other than the familiar yellow and black wasps that form large and often troublesome societies in late summer and early autumn. For convenience I use it to embrace the sand-wasps or digger-wasps (*Fossores*) of every description, as well as the solitary mud-wasps (*Odynerus*) and the social-wasps known to everybody. Members of these two last-named groups are easily recognisable by their habit of folding their fore-wing along its entire length when in repose: the hinder half of the outstretched wing is, in repose, doubled underneath the front half, so that the front wing then exposes only half of its full width. For this reason the mud-wasps and the social-wasps are grouped together under the name *Diploptera* (double-winged). Since, however, it is not our object to present a treatise on the anatomical structure, but rather an account of the interesting and fascinating ways of merely a few of these remarkable insects, we will now pass on to consider the habits of some of the "diggers."

CHAPTER II

FOSSORES OR DIGGER-WASPS: POMPILID SECTION

FROM an evolutionary standpoint the insects of this section are the lowest of those with which we are concerned; and it is interesting to note here and there among their members evidences of a tendency towards the higher and more complex conditions that now obtain among the most advanced of the social-wasps or in the honey-bee. The majority of the "diggers" are energetic, fussy, bustling insects inhabiting for the most part sandy districts, such as the Surrey heaths, or sand-dunes of our coasts.

They are notoriously fine-weather insects, and love a place in the sun above all else: in fact in dull weather it is useless to expect to see any of them. In size and colour they present a great diversity; some are no bigger than ants, others attain a length of about an inch; some are uniformly black all over, others black and red, others black and yellow like the social-wasps. They all provide their larvae with "fresh" animal food; some store caterpillars, some beetles, some small species of bees, some grasshoppers, some two-winged flies, some spiders for the nourishment of their offspring. The prey of whatever kind it may be is not actually killed but is merely paralysed; so that it remains fresh and virtually alive until the grub of the digger devours it. The parent digger secures this inert condition of her victims by skilfully stinging one or more of the chief nerve centres and rendering them inoperative by her poison. The species are all "solitary," that is to say each nest, or rather burrow, is the work of but one female; and she alone is responsible for the welfare of the young. The various kinds make their burrows in all sorts of places; many dig holes in the earth, preferring a light soil for obvious reasons, others dig galleries into wooden posts or decaying tree trunks, or into bramble stems, straws and similar objects. So far as my own experience goes very few, not even the largest, have a sting sufficiently powerful to penetrate the human skin, nor do they ever make any attempt to attack the observer, be he never so aggressive.

Some of the most interesting and most easy of observation among the digger-wasps are those known to science as the *Pompilids*: we have about thirty species belonging to this family, and fifteen of these are included in the genus *Pompilus* itself. The insects are rarely to be seen except in bright sunshine, for they hide underground or crouch motionless and difficult of detection when the sunshine passes away: the mere shadow of a passing cloud is quite enough to quench their activity for the time being. The majority are black, or black and red in colour; a few are black with creamy white spots. They all have long, wiry legs, whose first joint (i.e. that nearest to the body) or "coxa" is very large; the coxae of the second pair of legs actually meet each other in the mid-line underneath. The enlarged and closely approximated coxae are of great value to the insects when excavating their burrows. These nest-tunnels are often driven to a depth of several inches, and all the soil to be removed is brought to the surface by means of the enlarged and rather flattened coxae which act like hoe-heads beneath the body and drag the soil along the floor of the tunnel as the animal backs up to daylight from the dark recesses of her gallery: the close apposition of the middle pair of coxae ensures that very little soil slips between to be left behind. Arrived at the entrance the Pompilid scatters her load by vigorous kicking with the hind legs; so vigorous indeed that often there may be seen a fine jet of sand streaming, like water from a syringe, to a height of an inch or so, out into the air from the mouth of the burrow.

The burrow completed, the wasp catches a few spiders, each species usually adhering to some one particular kind of spider, paralyses them and conveys them underground where they are destined to serve as food for the grub which emerges from the egg that she attaches to one of her victims.

One of the most abundant of these Pompilids, known as *Pompilus plumbeus*, occurs on nearly all our sandy coasts, and not infrequently at inland places, from June until autumn sets in. The female is black and about a quarter of an inch in length, the male rather smaller and grey in consequence of the fine hairs with which his black body is clad.

I have studied the habits of these fascinating little wasps both on the sand-dunes to the north of Yarmouth, and on those of Braunton Burrows in North Devon. Whenever the sun shone brightly these active creatures were to be seen scurrying restlessly about with all the airs of a busy man in a desperate hurry. Seldom flying further than a few feet they transact their affairs on *terra firma*; their long wiry legs doing more than their fair share of work. Now and again a pause ensues in the bustling career, and the little creature digs furiously for a few seconds as though desirous of making a burrow; occasionally she halts and basks with outspread legs upon the warm sunlit sand. Some perseverance is needed if one wishes to witness the complete drama enacted; many a *P. plumbeus* will elude one's vigilance and with an extra rapid movement escape from observation. Let me narrate the events of one of my successful trackings: —having selected a specimen whose business-like demeanour seemed to promise a reward I followed on hands and knees her wild career through the tangled marram grass and over bare tracts of sands until at length, after much crawling and more perspiring, I tracked her to her burrow. As she neared home her excitement passed all limits, she leapt repeatedly a few inches in the air, and at last rushed headlong into the burrow; but no sooner in than out again and racing rapidly in all directions round the hole, taking, as I believe, the exact bearings of the spot so as to assist her in returning without waste of time. The survey completed, she dashed off: I rested by the burrow to await events. After about twenty minutes back she came, just put her head into the burrow, and was off again. I now followed her and found that about five feet away she had a spider which she had paralysed with her sting. She went straight from burrow to spider without any hesitation as to direction, seized it by one leg, and walking backwards dragged her victim a foot or so nearer the hole: then once again she must satisfy her anxiety that all is right at home, and back she scampers to inspect her front door; off again to the spider, brings him another foot or so on the journey; again, is home quite safe? back once more to her victim; but she had left him on a sloping bit of sand, and he had rolled helplessly down a few inches; so "when she got there the cupboard was bare." I laughed outright to see the mute astonishment depicted in this ferocious little huntress when she did not find her spider where she expected :—she turned round, looked in every direction, waved her antennae to and fro, as though to say "Surely I left him here; this certainly is the place." However, she wasted no time, but made a cast round the spot and soon recovered her treasured victim and resumed her task of alternately dragging it along and inspecting her burrow. At last she had the spider at the door of her den, and then entering backwards she dragged it down after her very rapidly and disappeared. After waiting vainly for half an hour in hopes of her return, I rose to go, intending to return later in the day; but in getting up I loosened some of the fine wind-blown sand, and a petty avalanche swept down smothering the mouth of the burrow. With a grass-stem I did my clumsy best to re-open the hole, and departed. On my return I found that the sand-wasp had disdained my proffered assistance and had opened a new hole just below my own unskilled attempt.

This little accident gives one possible clue to the incessant anxiety exhibited as to the welfare of the burrow. The surface of the sand is so loose that a very slight or even no apparent disturbance will set quantities of it sliding; so that it must frequently occur that the mouths of the wasps' burrows get smothered. Hence it becomes of prime importance to make sure that the hole remains open when the wasp is actually engaged in laying in supplies for her offspring. The opening up of the fresh hole is probably also a matter of daily occurrence; for at the end of any hot day very few holes are to be seen, and promiscuous digging into sand which had no holes visible frequently resulted in exhuming numerous sand-wasps which were buried for the night, and which would be constrained to make for themselves an exit next morning. Further it is apparently easier to make a clean hole in wind-blown sand by working from within outwards than in the reverse direction; for this species was never observed to succeed in digging a burrow at any of the numerous attempts upon the surface: doubtless also the slight coherence possessed by the surface when moist with the morning dew makes the task of constructing a neat and sharply defined entrance somewhat simpler.

To return to the individual we have followed: a careful sifting of the sand yielded two paralysed spiders, each with one egg laid upon it and fastened in exactly the same place, viz. on the front end of the upper side of the spider's abdomen and slightly to the right of the middle line. I suspect that even if the spider recovers from the effects of the sting it is unable to detach the egg, or grub resulting therefrom, from this position. Numerous exhumations of spiders always brought to light the same species (*Lycosa picta*) and always with the wasp's egg or grub in the same inaccessible position. The constancy of this one species of spider as the object of the attentions of *P. plumbeus* raises the question —by what sense, that of sight or of smell, does the wasp find and follow her prey? The behaviour of the wasp when actually on the trail certainly points to her following a scent; her movements instantly recall those of a hound, while the quick quivering touch of her antennae upon the ground suggests that she is gaining information by these, her olfactory organs. On the other hand it is noteworthy that of the two spiders which alone were abundant upon the sand-dunes, *Lycosa picta* is by far the more conspicuous, being of a dark grey colour with a few black markings, which show up plainly, especially

FOSSORES OR DIGGER-WASPS

when in motion, against the pale brown sand. The other (*Philodromus fallax*), though quite as abundant, is extremely hard to detect for its colour exactly matches the surface of the sand; and even when the spider is moving, the eye often fails to locate it though the observer is vaguely conscious that something within the field of vision has moved. It is thus difficult to avoid the conclusion that the wasp uses her eyes in addition to her olfactory organs in her quest, and that the colouration of *Ph. fallax* is not without its value. Every individual *P. plumbeus* does not behave in precisely the same fashion; there are degrees of intelligence among them. I have observed some which, like the above, took no pains to conceal their victim when they left it during their hasty visits to headquarters; while others that I have watched invariably buried their spider in the sand before quitting it. In one instance this temporary burial was always carried out near some conspicuous object, such as a small tuft of grass, a dead reed, or other similar landmark. The association in the wasp's mind between the site of her *cache* and the nature of the landmark was very evident; for if a tuft were the landmark then other tufts were often inspected during the search for the right one; also when I moved a dead reed near which the spider had been buried the wasp was much bewildered and divided in mind between the actual burial spot and the reed now some eight inches away, running distractedly to and fro between them. When the reed was replaced in its original site her excitement was intense, and the spider was promptly exhumed without further delay. In some cases the possession of a spider appears to upset the mental balance, or at any rate the sense and memory of direction; for one stupid individual who was always able to run home without her spider invariably lost

Fig. 1. *Pompilus viaticus*
(about twice natural size)

her way and failed to reach the desired goal when dragging her prey. Regret was impossible when another and presumably a cleverer *P. plumbeus* robbed her of her spider, buried the booty and eventually conveyed it successfully to her own nursery-larder.

On inland sandy heaths two of the largest Pompilids to be found are *P. viaticus* and *P. rufipes*.

II] FOSSORES OR DIGGER-WASPS 15

Both species provision their burrows with spiders and take great pains both in preventing the marauding ants, which abound on such heaths, from stealing their prey and also in concealing the position of the burrow when once it is properly stocked and the egg laid. *P. viaticus* is a handsome red and black wasp nearly half an inch long: it appears to love sandy patches, such as those thrown up by rabbits, in the midst of heather. The spider which is hunted by this sand-wasp is a large fat-bodied creature several times heavier than the huntress herself. The burrow is always made of such dimensions as to admit the plump body of the victim, and appears unduly wide in comparison with the slender abdomen of *P. viaticus*. This wasp appears to indulge in fencing matches by way of practice for her encounters with the formidable spider which she ventures to attack. Frequently two females—note this is no love-dance for both are females—may be seen to settle down face to face on a patch of sand, to move round and round as though searching for an opportunity, to lean over first on one and then on the other side, and from time to time to lash round with their wonderfully flexible abdomen as though delivering a stab from the venomous sting. In their actual encounters with the spiders it behoves them ever to face the enemy, for the poison fangs of the spider are situated at the head end, and it is thus from that quarter that danger threatens. The wasp on the contrary carries her weapon at the tip of her tail, and needs to be expert in at once keeping her eye on the foe and at the same time delivering an attack from her rear armament. Of course the poor spider is heavily handicapped by the absence of wings which enable the wasp to move and "make circles" round her less agile opponent. To equalise the chances on one occasion I put an uninjured spider and a female *P. viaticus* together in a glass-topped box which allowed plenty of floor space for the "ring," but was too shallow to allow the wasp to use her wings freely. The duel was of brief duration: wasp and spider stood facing each other tense on tip-toe: each plainly realised that a crisis had arrived: the wasp made a weak feint with her abdomen: the spider lunged quickly forward and for an instant touched the wasp: a spasm passed over the wasp and she fell over on her side dead. The spider had got "home" with the poison fangs and for once *P. viaticus* had to acknowledge defeat.

Under normal conditions the vanquished spider is seized by one leg and dragged helpless through the rough grass and heather towards the burrow. The wasp makes several visits of inspection to her home during the process of transport, but these seem to be for the purpose of maintaining rights of ownership and of ejecting parasitic flies and other insects that are ready to appropriate a tenantless hole or to play the cuckoo. On such occasions the spider is not left upon the ground exposed to the attacks of ants, but with much labour and struggling on the part of the wasp is dragged up a heather stem and carefully lodged in the fork made by two twigs, thence to be reclaimed after but very little search by the unforgetful owner.

The care with which the wasp conceals the position of the stocked and completed nursery-larder is very pretty to witness. First she scratches sand over the hole from all directions, until, to the human eye at any rate, all trace of the burrow has been obliterated; then, as though not satisfied that the surface presents no appearance of having been disturbed, she will gather dead heather bells, a few pine-needles, a pellet or two of rabbit's dung and scatter them in haphazard fashion over the place where her treasure lies hid. And what further interest takes she in this spot over which for the sake of her child she has expended so much care and labour? Absolutely none. To the best of our knowledge she never comes near the place again except by accident, but sets to work to repeat the whole performance whenever the conditions are favourable to her, and again exhibits the same heartless indifference to the welfare of her offspring when once she has laid in the store of a few spiders. Maternal instincts have reached no great developments here.

Pompilus rufipes is rather smaller than *P. viaticus*, and is black with a few white spots on each side of the abdomen. The female is a very fussy individual and shy of being observed: like all her relations she provisions her nest with spiders. Different individuals adopt varying methods for the concealment of their completed burrow: one, whom I had watched stocking her nest, had driven the burrow horizontally just below the surface of the sand for a distance of about two and a half inches before striking obliquely downward. When all was ready for closure she worked for a long while to and fro on the side walls of her horizontal tunnel until eventually she so weakened them that the whole length of the roof collapsed simultaneously, leaving a shallow trough along the surface of the ground. The wasp seemed quite prepared for this effect, for so soon as she had shaken herself clear of the sand which fell on to her, she set to work to level up the trough by scratching sand into it from all directions in turn: there was thus formed a scratched patch of about three inches diameter around the site of her tunnel: the scratching had exposed the tips of a few buried pine-needles; these the wasp pulled up and laid flat on the ground, and then dragged a few rabbit's pellets on to the patch, and then at length was satisfied. A second individual on another occasion having sunk a nearly vertical shaft, filled it when stocked by scratching inside near the top and round the entry. A funnel-shaped depression was thus caused and this she filled up with very small pebbles and bits of dead grass which she brought in her jaws.

Actions such as these now instanced seem to point to an intelligence higher than mere instinct: the insects adapt their conduct to meet various contingencies: they appear to have a definite purpose in view, and that purpose they achieve in a fashion which makes it difficult to deny to them at least some glimmering of the reasoning faculty. Nevertheless among these Pompilids we find very little indication of the amazing cleverness of the social-wasps and bees, and comparatively little of the maternal solicitude which has in all probability led to the evolution of the complex societies and architectural skill of the higher Aculeates.

CHAPTER III

FOSSORES OR DIGGER-WASPS: SPHEGID SECTION

ANOTHER section of the Fossores, the Sphegidae, contains far more British representatives than does the Pompilid, upwards of ninety different species occurring in this country. The several genera are distinguished by such structural features as the number and arrangement of the "cells" of the front wings, the shape of the mandibles, the form of the waist (*petiole*), and so forth. Some of the largest and most common members of this section are the four species of the genus *Ammophila* (sand-lovers): all four are from three-quarters of an inch to one inch in length, are black and red in colour, and have rather long—two have very long and slender—waists. Like the Pompilids they dig holes in the ground; but instead of using their feet and the coxal joints of the legs as tools, they employ only their powerful jaws on this work, scraping and biting at the soil until a small pellet has been detached, then seizing it in the jaws and carrying it backwards up the burrow. On arriving at the surface they dart backwards a few inches and fling the pellet away. It is interesting to note that the coxal joints of the legs of these insects are not enlarged nor flattened like those of the Pompilids, and further that the Ammophilan method of digging with the mandibles is the same as that employed by those of the familiar social-wasps which build their nests below the ground surface.

The following observations made on *A. sabulosa*, the largest of our species, will serve to illustrate the habits of these insects. A specimen was detected in the act of commencing her burrow on the edge of a disused sandstone quarry: for nearly an hour did she work carrying out mouthful after mouthful of earth,

III] FOSSORES OR DIGGER-WASPS

and depositing the loads a few inches from the hole, sometimes on the right, sometimes on the left side; but invariably below the aperture so that none of the soil rolled back into the cavity she had made. The burrow was made some three inches deep, and while she was out of view *in profundis* the insect kept up an angry, rasping buzz which effectually scared away a fly of parasitic (cuckoo) habits who came to inspect

Fig. 2. *Ammophila sabulosa*
(slightly enlarged)

the mouth of the tunnel. All the Ammophilas have this habit of buzzing when down their burrow, and I have little doubt that the noise serves to warn off intruders on mischief bent, and so saves the lawful tenant the necessity of resorting to more forcible, but time-wasting methods. At length enough mining had been done, but finishing touches were needed at the entrance. The mouth was slightly enlarged so as to be funnel-shaped; and then was exhibited a delightful display of sagacity. The insect ran hurriedly round about her burrow, and with her jaws seized a tiny lump of earth, but at once dropped it: this she did several times, but at length found a small flat stone. This she grasped with evident satisfaction and at once ran off with it to her burrow. Now the purpose of the funnel-like entrance became clear: the stone went easily into the wide mouth of the funnel, but rested flat across the more narrow bottom, and served to prevent the pellets of earth and sand, with which she now quickly filled the entrance, from falling down the main shaft. Notwithstanding her haste, the work was done carefully; for more than once a pellet found to be too big or of the wrong shape when put upon the rest was at once removed and thrown away. The covering of the hole having been made nearly flush with the surrounding surface, the insect proceeded to nibble the sand above the entry, so that a shower of grains fell down over her work and made the spot look exactly like all the rest of the bank. To complete the concealment she cut a few dead grass roots and scattered them about the place. So perfectly was the burrow hidden that I deemed it advisable to mark the spot with a sprig of heather lest I should forget the exact site. The same need was evidently experienced by the insect herself, for she proceeded to take careful note of her surroundings, making at first short, subsequently longer, excursions in every direction round the burrow, and constantly assuring herself that she could find her way back. This exploration was gradually extended over a radius of several yards, in the course of which survey I myself and my entomological paraphernalia were subjected to a minute scrutiny. Eventually the wasp darted off, and for a long while was not in evidence. At length she returned astride of a large green caterpillar about half as long again as herself, and many times heavier. Quickly and straight to her burrow did she convey this paralysed burden: the covering of sand and the stone plug were removed and the caterpillar taken out of sight. Soon she reappeared, again carefully covered up the mouth of the burrow, and departed on a second hunt. This time it was a brown caterpillar of another species, but about the same size as the first victim. The same proceedings were gone through again, but after the hole had been covered up and carefully hidden, the little huntress sat resting in the sun for a few minutes. Then she flew off and began digging another burrow a few yards away. Exhumation of the two caterpillars revealed a large white egg firmly fastened upon the right side of the first victim.

A. sabulosa, be it noted, supplies her young one with two large caterpillars, and thereafter takes no further interest in its welfare. In this particular the behaviour of *A. campestris*, otherwise very similar, is interestingly different. This latter species captures small caterpillars, of such size that two would not be an adequate provision: she does not, however, complete the provisioning at the time when she constructs the burrow, but for several days returns bringing a few fresh caterpillars for her growing young every morning. The burrows of this species, unless they have but recently been made, invariably contain several freshly caught caterpillars, a few partly devoured, and the empty skins of a few others : and I have repeatedly observed *A. campestris* carry caterpillars into burrows which, when opened up, were found to contain large grubs that had evidently emerged from the egg several days before. Here then we find the beginning of the idea of a more permanent nest, of attachment to a particular spot, and of a more lasting maternal solicitude. It is interesting also to note that frequently a number of individuals of this species will nest side by side in the same small patch of ground. Such an association renders it conceivable that the neighbours may help one another, but here direct evidence is wanting and will be very difficult to obtain.

Another very common genus of Sphegid sand-wasps, though small both in the number of species and in the size of the individuals, is known as *Oxybelus* (the swift-darter). *O. uniglumis* is the species that is most abundant; it may be found in numbers on the blooms of stonecrop, gypsophila and many other shallow flowers during July and August. But it is when attending to the needs of her family rather than to her own that she becomes most interesting. The insect is about a quarter of an inch long, and is black with several white spots along the sides of the abdomen; both in size and appearance she is not at all unlike an ordinary house-fly, or any of the

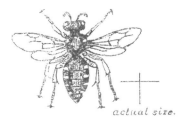

Fig. 3. *Oxybelus uniglumis*

two-winged flies that are so persistently annoying to human beings during the hot summer days; and like many of these last she has a covering of beautiful glittering silver hairs all over her face. Moreover in several of her habits she closely resembles the flies which frequent hot, sandy places; she sits basking on the bare stones in the way that the flies so often do, she flies in a very similar way, she turns her silvery face towards the observer, or towards a fly that happens to alight close by. All these resemblances to a fly appear to have a purpose. *O. uniglumis* provisions her nest with flies, and thanks to her disguise is able to pass and even play among the flies without arousing suspicion. She affords us one of the very few examples of what is termed "aggressive mimicry." Her main object in life is to secure as many flies as possible with the minimum of trouble. Her burrow is about three inches deep, and is always placed close by or even in a patch of bare soil where flies may bask; but she disdains the "sitting" shot, and invariably catches her prey in mid air, swooping on it like some tiny falcon. On leaving her burrow she covers the entrance by a few quick scratches at the sand and mounts up into the air to join in the throng of flies whose buzz overhead may be heard on nearly any hot, fine day. No time is wasted; in a few seconds she falls again to earth grasping a struggling fly in her strong legs: the quietus is quickly administered by the sting and the paralysed body dragged down the burrow. Six times in five minutes have I seen the same individual insect-hawk successfully repeat the whole of the above tragedy; and when the completed burrow was opened up there lay sixteen helpless flies intended as food for the grub which should have emerged from the single egg placed among them.

This little wasp has a very peculiar method of

conveying her prey into the burrow: the majority of the Fossors go backwards when dragging their victim into the hole; but *Oxybelus* always goes in head first, and instead of using jaws and legs to hold her prey, she pierces it with her sting and so gains a secure attachment. The paralysed fly is always thrown on its back; its head is directed in the same way as the head and lies beneath the thorax of its captor; its thorax underlies the abdomen, and its abdomen projects out beyond the tail of the little wasp. The sting is driven in on the underside of the last part of the thorax. I am aware that this account is at variance with that given by such careful observers as Mr and Mrs Peckham for *O. quadrinotatus*. But I have had *O. uniglumis* and her attached prey in a glass tube repeatedly, and examined the two carefully with a strong magnifying glass, and am convinced that though when on the wing the wasp holds the fly with her legs, and it is noteworthy that the adhesive pad (*pulvillus*) between her claws is extraordinarily large and presumably powerful, yet once the *coup de grâce* has been administered, it is the sting that keeps the hold. The following incident confirms this view: during the brief absence of the little wasp some sand had fallen across the mouth of her burrow, so that there was more than usual to be removed in order to clear the entrance: the wasp scratched at the sand with her front legs for a few moments, then, in order to afford a clear path along which to kick the loosened sand, she elevated her abdomen over her thorax, *holding the fly up in the air impaled on her sting at the tip*, and with her hind legs scattered the little heap that had been gathered by the front feet. I watched the whole of this manœuvre through a strong lens.

Among our British Sphegidae the genus *Crabro* is represented by the greatest number of species, and may therefore claim some notice in these pages. The insects comprised in this genus differ widely from one another in superficial appearance, some being entirely black, while others are conspicuously banded with yellow and a few with red: in size too they vary from nearly three-quarters of an inch to less than a quarter of an inch in length. They may, however, be recognised without much difficulty by the peculiar arrangement of the *cells* into which the forewings are mapped out by the *veins* or *nervures*, if two other fairly obvious features present in two small genera which share this peculiarity be taken into account. In any of the Hymenoptera dealt with in this volume it will be seen that there are in the forewing four "nervures" which start from the attachment of the wing to the body and run out horizontally towards the tip of the wing: the foremost of these "nervures" forms the front margin of the outstretched wing and is known as the *costal nervure*: very close behind it and almost

FOSSORES OR DIGGER-WASPS

parallel with it is the *post-costal nervure*: both these nervures end in a dark thickening, known as the *stigma*, situated on the front margin at about

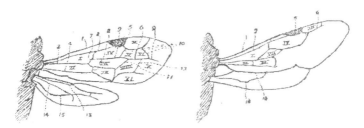

Fig. 4. On left, wings of a Hymenopteron showing the full arrangement of nervures and cells. On right, wings of a *Crabro* (both much enlarged).

Anterior wing:

1. Costal nervure.
2. Post-costal nervure.
3. Median ,,
4. Posterior ,,
5. Stigma.
6. Marginal ,,
7. Upper basal nervure.
8. Lower basal ,,
9. Cubital ,,
10. Submarginal nervures.
11. 1st recurrent ,,
12. 2nd recurrent nervure.
I. Upper basal cell.
II. Lower basal cell.
III. Marginal ,,
IV. 1st submarginal cell.
V. 2nd submarginal cell.
VI. 3rd ,, ,,
VII. 1st discoidal ,,
VIII. 3rd discoidal ,,
IX. 2nd discoidal ,,
X. 1st apical ,,
XI. 2nd apical ,,

Posterior wing:

13. Anterior nervure.
14. Median nervure.
15. Posterior ,,

two-thirds of its length from the attachment to the body. From the stigma there starts a nervure which at first sweeps backwards (in the outstretched wing) and then forward again and outward up to the front margin so as to enclose an area termed the *marginal cell*. The two other horizontal nervures, taken in order, are respectively the *median* and the *posterior*. In order to explain the peculiarity of the forewing of a *Crabro* we must crave the indulgence of the reader a little further, and mention a few of the nervures which run across the wing in a more or less transverse direction—the remainder of these shown for the sake of completeness in fig. 4 need not here concern us. The first of these transverse nervures, starting from the body, occurs at about a third of the length of the wing and extends from the postcostal across the median to the posterior nervure: it is known as the *basal nervure*, and is not straight but more or less zigzag with distinct angles along its course. It will be seen that this basal nervure marks out two "cells"; one, the *upper basal cell*, between the post-costal and median nervures; the other, the *lower basal cell*, between the median and the posterior nervures. The basal nervure thus forms the outer boundary of both these cells. From about the middle of both the upper and the lower part of the basal nervure, i.e. from about the centre of the outer boundary of both the upper and lower basal cells, there starts out horizontally a nervure which proceeds towards the tip of the wing. That from the middle of the outer edge of the upper basal need alone concern us : it is termed the *cubital nervure.* As this cubital nervure passes outwards it is joined to the curved nervure which encloses the marginal cell by one, two, or three short transverse nervures, which thus enclose one, two, or three areas, the *submarginal cells,* immediately behind the marginal. Now the number and shape of these submarginal cells are found to afford very useful characters in discriminating between the various groups of these aculeate Hymenoptera. For example, in *Ammophila* there are three submarginals, and while in *Ammophila campestris* the third is united to the marginal by a single nervure (*petiole*) which branches like an inverted letter Y [⋏], in *A. sabulosa* the petiole is absent and each of the two limbs of the ⋏ is joined independently to the curved nervure enclosing the marginal cell: in certain other genera there are only two submarginals, one of which is in some provided with a petiole, while in others no such structure is present. But in *Crabro* there is but one marginal cell in each forewing. Only two other genera of *Sphegidae*, namely *Oxybelus* and *Entomognathus*, have but one submarginal; and these two are easily distinguished from *Crabro*; *Oxybelus* by the presence of a sharp spike on the back of the hinder part of the thorax, and *Entomognathus* by the hairy covering with which its eyes are clad. As a rule the *Crabros* have large, squarish heads and very large eyes; and in some the front of the face is covered with very beautiful silvery or golden hairs which glitter like polished metal in the sunlight. The form of the body is very varied: in a few there is a long and slender waist connecting the swollen portion of the abdomen with the thorax; while in others the waist is quite short, or perhaps hardly distinguishable at all. But it is in the strange forms assumed by the legs, especially by the anterior pair of the males of some species that this genus is most remarkable. Thus in *Crabro tibialis* the tibiae of the hind legs are extraordinarily swollen so as to resemble little clubs; and the metatarsal joints of the front legs of the male are very long and widely dilated. Again, in *C. cetratus* we find that the outer edge of the tibia and metatarsus of the front leg of the male is spread out in a thin membranous expansion which carries a fine fringe of bristles; while in *C. gonager*, a small, polished, black species, there are somewhat similar enlargements on the front tibiae, and on the front metatarsi large outstanding shields which are coloured white with three black spots; the succeeding tarsal joints are also widely spread out. In *C. palmarius* also the male has conspicuous shields on the tibia and metatarsus. Further eccentricities of structure occur in

III] FOSSORES OR DIGGER-WASPS 33

the legs of such species as *C. cribrarius* and *C. peltarius*. In the male of the former the femora of the front legs are each drawn out posteriorly into a large, twisted and five-sided process, and carry between this remarkable structure and the attachment to the body a sharp prong; while the corresponding part of the front leg of the latter is adorned with a flat, polished, yellow outgrowth which has at its base a very slender spike. The front legs of the male of *C. peltarius* are yet further singular in that the inner of the two claws is flattened and spread out at its base into a thin plate which is drawn out at its tip into a narrow twisted spine. Curious modifications of more or less similar form occur also in the legs of the males of *C. scutellatus, C. interruptus* and *C. clypeatus*; but these need not be specified in detail.

Enough has been said to call attention to these peculiarities that occur with such frequency within the limits of this one genus; and we trust that the reader has begun to wonder, as we ourselves still do, what purpose and function is served by these odd and often ungainly developments and departures from the ordinary type of structure. We regret that we are entirely unable to give any information on these points with regard to any one of the species concerned. So far as we are aware no knowledge has ever been obtained as to the use or uses of any of these structures. It is difficult, to a believer in

Natural Selection it is impossible, to imagine that all these singular contrivances are mere purposeless accidents playing no part in the general economy of the species. The fact that they are confined to the male renders it certain that they are not concerned with procuring food or storing larders for the benefit of the offspring, and makes it not improbable that they are secondary sexual characters which perform some useful function, perhaps in the act of mating; but of their exact use we are absolutely ignorant. It is to be hoped that perhaps some few of those into whose hands this volume may come will be stimulated by this frank confession of ignorance to make a patient study of these interesting insects, and eventually discover what part is played by these singular excrescences to which we have now called attention. It would not be difficult to draw up a list of almost equal extent dealing with curious modifications in the shape of the antennae, or at least of portions of them; and here again we should be compelled to admit total ignorance. But to ascertain the uses of the several modifications in this case would necessitate expert knowledge and training in the methods and technique of microscopy.

In their general habits these Sphegids are not unlike the other Fossors, for the females stock the burrows in which they deposit their eggs with two-winged flies and other insects. As might be expected

III] FOSSORES OR DIGGER-WASPS 35

in a genus so extensively represented in number of species, there is considerable variety in the choice of nesting sites. Some, such as the common little species *C. wesmaeli*, a shining black and yellow insect with a minute red tip to the abdomen, and the larger *C. cribrarius*, excavate tunnels in sandy soil; some, such as the equally abundant yellow and black form, *C. quadrimaculatus*, burrow into rotten wood; while others succeed in boring into the comparatively sound timber of old gate posts, wooden fences and the like; and a large number select the stems of brambles and briars, boring into the pithy centres, as their place of nesting. From the fact that their prey is entirely composed of insects, they may without hesitation be regarded as useful to mankind. I have several times seen *C. clavipes* catching the small flies that in some states of the weather accumulate in large numbers upon the window panes of sunlit rooms. *C. leucostomus* provisions its nest with a bright green fly (*Chrysomyia polita*). F. Smith states that *C. podagricus* lays up stores of a certain small species of gnat. Some of the larger and more powerful species, e.g. *C. dimidiatus*, aim at larger game and succeed in capturing large blue-bottle flies; while many of the smaller kinds store up midges and other troublesome little insects, or collect plant-lice (Aphides) from our garden plants. The full-grown insects support themselves upon vegetable products, and may often be seen visiting various umbelliferous plants and others whose nectar is placed within reach of their short tongues.

CHAPTER IV

DIPLOPTERA (DOUBLE-WINGED WASPS)

As already stated the members of this, the highest of the wasp families, are characterised by the longitudinal fold into which the front wing is thrown when at rest. Of the two sections comprising the family the Eumenidae or mud-wasps are solitary, while the Vespidae or social-wasps form the great communities with which every one is more or less acquainted. There are also structural differences by which members of the two sections can be distinguished: the solitary forms have a longitudinal furrow running along the outer face of their mandibles, and have a small "tooth" near the apex of their claws; whereas the mandibles of the social-wasps have no such furrow, nor are their claws toothed. We have only two genera of mud-wasps represented in this country, viz. *Eumenes* and *Odynerus*. Both are narrow bodied, and yellow and black in colour. The former is characterised by an elongated "waist." Its sole British representative, *E. coarctata*, makes little globular nests of mud which it attaches to twigs of heather; and, like all the solitary Diploptera, provisions them with small caterpillars. We have fifteen species of *Odynerus*, all very much alike in general appearance, though exhibiting considerable differences in their nesting habits. *O. spinipes* makes her nest in banks, generally of rather stiff soil, and constructs with pellets of mud a remarkable curved projecting entrance which juts out from the face of the bank like

Fig. 5. *Odynerus parietum*
(about twice natural size)

a downwardly directed spout. The wasp has to climb up this spout before reaching the down-shaft leading to her larder. Some of the other species of *Odynerus* make their nests in bramble stems and line the tunnels with a coating of fine sand. Others construct mud cells inside door locks, in the slots for window-catches, in crevices in masonry, or in any angle afforded by walls, window-ledges and so forth.

It was in one of these last situations that I had the opportunity of closely observing a specimen,

O. parietum. The little wasp had selected the angle formed between the frame and the ledge of my study window as a suitable spot for her operations. I discovered her when she had nearly completed the first of the series of cells. A few more loads of mud were brought to complete the horizontal sides and roof; then the wasp backed into the cell and, as was afterwards discovered, laid one egg, suspending it by a silken thread to the roof of the cell. Over a dozen small green caterpillars were then in quick succession brought paralysed to the cell and rammed in by vigorous butts from the wasp's head. At length the vertical end wall was added to close the cell and to furnish the back of the second cell of the series. Knowing that in all probability a second cell would be built at once, in the wasp's absence I carefully opened the completed cell and laid the caterpillars and her egg, which was found as already described, beside the ravaged nest on the window-ledge. On her return the insect manifested great consternation at the havoc wrought to her handiwork. However, nothing daunted she hastily gathered the caterpillars together and once again rammed them into the cell, repaired the damages to the walls, and closed the end. But she took no notice whatever of the egg, nor had she, as was ascertained by a later inspection, laid another egg. Her instinct led her right in so far as she realised that the caterpillars should be sealed up in the cell, but her egg was a new and unfamiliar object: probably she had never seen her own before, and it was certainly asking something entirely outside her previous experience to expect her to place an egg in the cell by means of her jaws and feet. Here then instinct failed; and the failure revealed the utter absence of reason from her other actions in these novel circumstances.

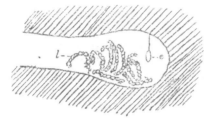

Fig. 6. Nest of *Odynerus*
e = egg of *Odynerus* suspended by silk thread
l = caterpillars stored by *Odynerus*

All these mud-wasps appear to behave in the same way in suspending their egg by a thread to the roof of the cell. The egg is at the far end of the cell, and the caterpillar first brought in lies closest to it: the others lie in order between this and the cell-door. When the grub first hatches from the egg it is very small and weak, and might easily receive a fatal injury from the semi-paralysed caterpillars, for they still retain some power of movement—they can use their mandibles and "kick" with the hinder part of their body. Hence it is to the advantage of the grub to retain its hold upon the egg-shell and make its first meals from above, safely out of reach of any too vigorous protest on the part of its victims. To enable the little larva to reach caterpillars rather further away than the first devoured, the egg-shell splits up to a sort of ribbon, and thus increases the length of the suspensory thread of which it forms a continuation. By the time that the larva has eaten all the caterpillars within reach of its lengthened tether it has so increased in size and strength that it can now venture to loose its hold on its safety rope and boldly eat its way forward among the steadily decreasing mass of caterpillars. Eventually it pupates and completes its metamorphosis.

The species of *Odynerus* are very subject to the attacks of "cuckoos" belonging to the *Chrysididae*, the jewel- or ruby-wasps as they are popularly termed; and also of certain dipterous flies of similar habits. The careless manner in which all the *Odyneri* leave the mouth of their cell or burrow wide open during their excursions in search of mud or of caterpillars renders them easy victims to this imposition.

Our social-wasps, including the hornet, number

seven species, all included in the genus *Vespa*. Their nests, though founded by a single individual, are the work of a vast host composed of the daughters of the foundress "queen." In them are to be found during the later part of the summer (1) perfect, fertile females, or "queens," of whom one is the originator of the nest and mother of the entire colony, while the remainder are young "queens" whose function it is to found similar nests in the following season: (2) imperfect, sterile females, or "workers," on whom falls the brunt of building the nest, providing the grubs with food, and preparing cells once tenanted for use a second and a third time: (3) drones, or male wasps whose sole duty is to impregnate young "queens" before severe weather sets in.

The hornet (*V. crabro*) is much larger than any of the others, being fully an inch in length: it is distinguished by its colour, which is reddish brown and yellow; whereas that of all our other *Vespae* is black and yellow: it usually constructs its nest in hollow trees, or in the roof of a building; but I have seen one in the side of a bank of stiff grass-covered clay. This formidable insect is now decidedly rare.

Of the other six species, *V. vulgaris*, *V. germanica*, *V. rufa*, generally build their nests under ground, though sometimes they resort to a roof, a stone wall, the thatch of a rick and other similar situations. On the other hand, *V. norwegica* invariably and *V. sylvestris* nearly always hang their nests from the branches of shrubs or trees. The remaining species, *V. austriaca*, is by some regarded as a distinct species, by some as the ancestral form of *V. rufa*, and by others as a "cuckoo-wasp" that patronises the nests of *V. rufa*. This is not the occasion for a description of the features by which these species may severally be distinguished; we may state, however, that *workers* of the two arboreal species have a yellow stripe on the front of the first joint ("scape" or stalk) of the antenna, whereas the workers of all the terrestrial species have the "scape" entirely black. This diagnostic character is not applicable to the "queens" or to the drones. "Queens" of all species are recognisable by their greater size; they are only to be found on the wing in spring and early summer, and again, though much less in evidence, in early autumn. Any *Vespa* found flying about before the end of May is certain to be a "queen." The drones do not appear till August and September; they have longer bodies, longer antennae (13-jointed, instead of 12-jointed as in females), and they are not armed with a sting. They frequent the flowers of wild parsnip and other umbelliferous plants, and also of the ivy, provided that bad weather has not cut short their career before this plant is in bloom, and probably of many others; and may be picked off the blossoms with the fingers with impunity—the great length of the antennae is so noticeable that a disagreeable mistake is not likely to occur when once the captor has seen a drone.

The structure and general economy of a wasp's nest will be best understood from a narration of the history of this wonderful piece of insect architecture

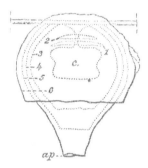

Fig. 7. Diagram of wasp's nest at an early stage

ap. entrance to nest. *c.* comb. 1, 2. 1st and 2nd coverings now largely removed to make room for growing comb. 3, 4, 5. corresponding coverings still complete. 6. 6th covering as yet not finished.

ab initio. The young "queens" having mated with drones in the previous autumn retire into winter quarters. Any snug corner may serve as a suitable spot for hibernation, e.g. the folds of a curtain, the corner of an outhouse, thatch, thick and fairly dry moss. Having selected her resting place the "queen" grasps a shred of fabric, be it curtain or straw or what not, between her mandibles; folds her wings beneath her body and becomes dormant; the claws on her feet take little, if any part, in holding her in position. In this state she is capable of resisting a surprisingly low temperature. I have known a hibernating "queen" exposed to a temperature of 10° F. (22° F. below freezing-point) without any ill effects. On the other hand unseasonable warmth awakes them and they are then liable to disaster. I have seen "queens" on the wing out of doors on December 26th and on February 7th. Hence a severe winter in all probability is favourable to wasps and likely to be followed by an abundance of "workers" in the following summer; whereas if the winter be mild with a few intermittent "snaps" of hard weather many prematurely awakened "queens" are killed.

The first business of the "queen" duly aroused by the warmer days of spring is to find a suitable spot for her nest. In April and May she is often to be seen prospecting along hedge-banks and walls for an eligible building-site. If she be a ground-builder, and we will consider one of these species as they are the most common, she will probably pitch upon an old mouse-hole, or mole-run, as the favoured spot. The surroundings of the site are carefully surveyed by the "queen" in order that she may know her bearings and be able to return home without loss of time. She may be seen flying to and fro in ever widening oscillations over the hole until all the neighbouring objects are impressed upon her memory. Before actually beginning to build she excavates with her jaws a small chamber at the end of the tunnel, carrying the loosened soil out on to the surface. She then proceeds to gather material for the construction of her nest. This she obtains by rasping off with her jaws the weathered surface of wooden posts, palings, etc.; it is very seldom that rotten wood is used either by the "queen" or by the workers. Nearly any oak fence in the open country bears upon it during the summer time hundreds of marks as if it had been lightly scraped with the finger nail: these marks are made by the jaws of wasps. Having collected a small pellet of wood-fibre moistened with saliva from her mouth, the "queen" carries it off and applies it as a thin layer of "wasp-paper" to the top of the cavity which is to hold the nest, usually attaching it to one or more small roots capable of sustaining the weight of several pounds. By repetitions of this process at length a disc is formed, and from the centre of this is hung a narrow stalk which widens out at its lower end to form the hexagonal outlines of the first four cells. These four are arranged in the form of a cross which forms the centre of the comb of cells constructed by the "queen." This comb contains about two dozen cells; they are all closed above but open

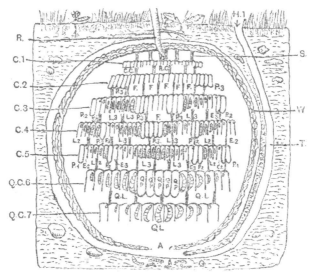

Fig. 8. Diagram of wasp's nest towards the end of summer
A., entrance to nest. C.1—C.5, first to fifth combs. C.C., cells closed with wasp-paper. E.2, E.3, cells containing eggs for the second and third time respectively. F., cells that have been thrice tenanted by grubs, and are now discarded for breeding purposes. G., larvae of a dipterous fly acting as scavengers of the *débris* that is ejected from the nest. H.1, hole in the ground by which wasps approach A. L.2, L.3, cells containing larvae in various stages of growth for the second and third times respectively. P.1, P.2, P.3, "capped" cells containing pupae (or *imagines* ready to emerge) for the first, second and third times respectively. Q.C.6, Q.C.7, sixth and seventh combs containing "royal" (queen) cells only. Q.L., Q.P., cells containing larvae and pupae respectively which are destined to produce young queens. R., root to which nest is suspended. R.C., cells used thrice and now reduced in height. S., pillars of wasp-paper suspending comb to comb. T., tunnel from H.1 to A. W., wrapping of wasp-paper.

below. As soon as each cell is outlined and provided with a low inverted hexagonal parapet, an egg is placed in it by the "queen" and fixed by a cement in the angle nearest to the centre of the comb. Meanwhile, in order to protect the comb from the drip of rain percolating through the soil—nothing is more fatal to wasp communities than heavy drenching rain in early summer—and to conserve such warmth as may be generated by her own small body, the "queen" has added first one, then a second and a third umbrella-shaped covering of wasp paper. These wrappings are attached to the foundation (top) and are eventually made into balloon-shaped envelopes completely concealing the comb, and leaving open only one small entrance at the bottom. The whole nest at this stage has a diameter of about one or one and a half inches: the "paper" made by the "queen" is of very fine texture and can at once be distinguished by its superior quality from that made later by the "workers."

In about a week, the exact length of time depending upon the temperature, a legless grub emerges from the egg. To avoid falling headlong out of the downturned cell the little creature retains a few of the hinder segments of the body within the remains of the egg-shell which is still glued to the cell-wall. The grubs possess jaws and are fed by their mother chiefly on the juices of animal food, such as caterpillars, Aphides, and flies. As they increase in size they periodically moult their skins; while the cells are made correspondingly deeper by additions of "paper" put on to their lower edges by the "queen." Subsequently the curved shape of the body of the grub and its general plumpness cause it to impinge firmly on both dorsal and ventral sides against the walls of the cell, and thus a secure hold is maintained.

When full grown the grub spins a silken cocoon within the cell, lining the entire cavity thinly but closing the bottom with a thicker and tougher floor which projects as a dome beyond the lower end. Shortly before the pupa (chrysalis) stage is reached, a black mass, composed of a sac containing the whole of the excrement accumulated during larval life, is ejected from the posterior end of the grub. This mass is not removed from the cell, but is flattened by the pressure of the larva against the cell-roof. Soon after the completion of the cocoon, the larva moults its skin and assumes the pupal condition. In this stage the insect possesses distinct head, thorax and abdomen and their respective appendages, viz. a pair of antennae, three pairs of jaws, three pairs of legs and two pairs of wings. These appendages are folded against the body and the whole pupa is of a semi-transparent white colour and very soft. During this period no food is taken, but very extensive internal changes are effected at the expense of the stores of fat accumulated within the body of the well-fed grub. Gradually the form and colour of the perfect wasp (*imago*) become visible through the delicate skin of the pupa which at length bursting open sets free from its shroud the fully-formed young worker-wasp.

The young *imago* bites through the floor of her cocoon with her mandibles and crawls out upon the under surface of the comb. The whole development from egg to imago occupies about a month or six weeks. At first the body of the young wasp is moist, and its colour dull; the wings too are not fully expanded nor firm: their final consistency, form and colouring are attained while the freshly emerged insect hangs, back downwards, from the comb. The excreta that have accumulated during the period of pupal quiescence are ejected at this stage. The front wings of the young wasp are not, at first, folded longitudinally: it is not until they have been extended that the ledge along their hinder margin becomes engaged on the hooks which lie along part of the front border of the hind wing: then, on the wings being brought into the position of repose, the attachment between the two wings is maintained and the hind part of the forewing folds mechanically under the front portion. One of the first acts of the wasp in full control of her limbs is to clean herself with the brushes and bristles with which her legs are provided. The front legs possess a special arrangement for cleansing the antennae: if any living aculeate insect be observed for a few minutes, it is sure to be seen to pass its antennae through the brush-and-comb apparatus on these limbs. Her toilet completed, the young "worker" visits a few cells containing large larvae, touches their heads with her jaws, and obtains from each a drop of liquid which she at once swallows. For a day or two she remains within the nest, but relieves the "queen" of the labour of distributing food to the grubs; for when the "queen" brings a load of food to the nest, the young "worker" takes it from her and doles it out to the hungry larvae.

In about two days the young "worker" issues from the nest and sets about the task of collecting supplies for the grubs, and "wasp-paper" for the enlargement of the nest. When the "queen" has reared about a dozen helpers she delegates to them the entire care of the grubs and the work of building. Henceforth she herself remains within the nest and devotes all her energies to depositing eggs in the cells prepared for her.

The "workers" carry on the building begun by the unaided "queen." Fresh cells are added round the margin of the existing comb; the inner wrappings are cut away one after the other to make room for extension, and fresh coverings are added externally.

DIPLOPTERA 51

This process is repeated as the need arises, and entails much labour on the "workers": the material removed from inside is used anew for construction of cells or of the protecting envelopes. These last are very numerous and hold between them layers of air: they serve to prevent the escape of heat generated within the nest. The temperature inside a strong nest is between 85° F. and 90° F.

As the nest is made larger the soil is removed from the subterranean cavity in order to afford more space. "Workers" may be seen issuing one after another from the hole and carrying away load after load of soil in their jaws. The exact shape of the individual combs is normally circular; but should an immovable obstacle be encountered in the process of excavation, the shape of the combs and of the whole nest is adapted to meet the special circumstances.

When the first comb has reached a convenient size pillars of wasp-paper are built downwards from the angles of some of the more central cells. At the lower ends of these pillars a second tier of cells similar to the first is added; and so on with each successive tier as the strength of the colony demands. An interval just sufficient to allow the wasps to move freely about within the nest is left between each successive comb, and also between their margins and the innermost of the protective wrappings. In the central portions of the combs the cells are very regularly hexagonal, but near the margins they become irregular or nearly cylindrical. The number of combs may be, in a strong colony, as many as twelve, one below the other: about seven is however the usual number.

Each cell is not tenanted by a grub merely once; but so soon as the first imago vacates it, the "workers" clear away the remains of the cocoon, and a fresh egg is deposited by the "queen." The cast skins and mass of excreta are however left in the cell, firmly fastened to the roof. It is by counting these masses that the number of tenants of any given cell is determined; and it appears that no cell is used more than thrice. When this limit has been reached the walls of the cells are cut away so as greatly to reduce their depth, and frequently the stump of the cell is sealed over with a covering of "paper." In a very strong nest the topmost comb is sometimes completely cut away, its place being occupied by a thick mass of wrappings. The lowest combs, sometimes only one, sometimes as many as six, are composed of cells larger than those in the superjacent tiers: these "royal" cells are destined for the "queens" of the next generation. It appears to depend upon the amount of food supplied to a grub arising from any given fertilised egg whether the resulting female be a "queen" or a "worker." It sometimes happens in a favourable season when the "workers" are numerous and food supplies plentiful that worker larvae in consequence of their generous diet become fertile (parthenogenetically); their eggs however produce only drones. Hence the drones of any given community may be either sons or grandsons of the foundress "queen"; but in all cases they arise from unfertilised eggs: they are usually reared in ordinary, but sometimes also in "royal" cells.

If a comb from the centre of a nest in full activity be examined, it will be seen how that the "queen" in her egg-laying journeys travels round and round the comb in fairly regular circles. The marginal cells of such a comb will perhaps contain eggs but lately laid; the next band will contain young larvae, the next rather older, and the next full-grown larvae; then follow circles of "capped" cells (i.e. with the dome-shaped cocoon projecting) containing pupae or perhaps young wasps just ready to emerge: still nearer to the centre occurs a circular belt of cells just vacated; these are followed by cells that have been tidied by the workers and have lately received from the queen an egg for the second time; and then the whole series of stages is repeated again, circle after circle, as we pass towards the centre which is the oldest part of the comb.

In the course of a good season the population reared in a fair sized wasp's nest is very numerous. Janet counted 11,500 cells in a nest containing but seven combs; of these about 5000 had been used thrice and over 6000 twice only. It is thus quite probable that a large nest of ten or eleven combs might in the course of the summer have a population of some 50,000 or 60,000, and the vast majority of these the offspring of the original "queen."

Upon the outer wrappings of the nest there may generally be found numbers of small oval white specks. These are the eggs of a two-winged fly (*Pegomyia inanis*). The parent fly walks unmolested down the tunnel leading to the nest, and deposits her eggs at random upon the "paper" of the envelopes. When the maggots hatch out from these eggs they immediately drop to the ground below the nest. In the soil immediately under the entrance proper, which is always at the bottom of the nest, there is a considerable accumulation of organic matter. For the wasps are scrupulously clean in their habits, and always proceed to the entrance to void their excreta. Here hanging on to the edge of the envelopes of the nest they extend the abdomen vertically downwards and discharge their evacuations upon the ground below. It is in this miniature manure heap then that the maggots of the fly find their food; and it cannot be doubted that they play a useful part as scavengers to the wasp community above. The maggots of other flies may also be found within the nest itself: Janet saw the maggots of *Volucella inanis*, a fly which very closely mimics wasps in its colour pattern, moving over the combs and apparently devouring the larval excreta at the tops of the cells, in fact acting scavenger from door to door. Many species of both insects and of other small animals have from time to time been found in wasps' nests. Some are undoubtedly parasites, others thieves, others mere chance sojourners glad of a warm and dry shelter. Two of the parasites deserve further notice. A thread-worm was found by Kristof projecting from the body of a drone wasp; I myself have on several occasions made a similar observation: the fact that this parasite occurs, so far as we know, only in the drones raises a suspicion that some peculiar form of food is supplied to the drone grubs, but we have no evidence bearing on this question. The other is a remarkable beetle (*Rhipiphorus paradoxus*). The larva of this beetle is believed to leap upon the bodies of worker-wasps when they are gathering wood-fibre off the surface of timber. It is thus carried by the returning insects into their nest. It now eats its way into a wasp-grub and devours the less important tissues of its host: when it becomes of such a size as to threaten the life of its unfortunate victim, it passes out through the skin of the wasp-grub, plugging the wound with the skin which it itself moults as it issues, and now becomes an external parasite upon the same host. It refrains from killing the wasp-grub until the latter has spun its cocoon. Eventually the beetle larva completely devours the wasp-grub and accomplishes its own metamorphosis within the cocoon provided for it by its prey.

As the autumn advances the strength of a wasp community steadily dwindles, though individuals may linger on as late as early December. As food supplies fall short the "workers" hasten the end by themselves devouring any grubs and pupae that may still be within the cells. Eventually the entire population, with the exception of the young "queens" who, as already explained have long before gone into winter quarters, perishes, and the nest with its remaining contents falls to ruin under the attacks of mice, beetles, earwigs, woodlice, and the like.

It is perhaps advisable to enter a short plea on behalf of wasps. They are commonly regarded as an unmitigated nuisance. It should, however, be remembered that unless provoked they do not use their sting; so long as hasty movements are avoided it is possible to stand with impunity right against a wasp's nest and watch all that is going on. Also it is only when fruit is ripe that they do serious damage —granted that in a "wasp year" the loss inflicted may be very great: nevertheless in the earlier part of the season they are good friends to the gardener and fruit-grower, for they destroy enormous numbers of caterpillars, and especially of green-fly and black-fly,

IV] DIPLOPTERA 57

and of other harmful insects. Moreover in a small way they act as scavengers on the earth's surface, for they quickly strip the bones of small dead animals of all skin and flesh for the benefit of the grubs, and thus prevent carrion from becoming offensive.

CHAPTER V

BEES (ANTHOPHILA, OR FLOWER-LOVERS)

As in the predaceous Aculeates, so too among those which store up honey and pollen for their larvae, we find both solitary and social species. Many of the former have habits resembling those of the fossorial wasps: some dig holes in the ground, some in wood; some make their nests of bits of leaves, others of mud; while some advance towards the condition of the honey-bee in that they use both wax or other material secreted from their own bodies and the woolly coverings of the leaves and stems of certain plants (*Anthidium*) or moss and dry grass (species of *Bombus*) in the construction of their nests. It is indeed thought that the bees are descended from insects whose habits were those of the existing Fossors, but that they have progressed and undergone many modifications both in habits and in structure, particularly of their jaw apparatus and hind legs, since the time when they discovered that the food which suited the full-grown insect was equally capable of nourishing the larvae. Since again, the vegetable products, pollen and nectar, are far less liable than dead insects to undergo decomposition, storage of considerable quantities becomes possible; and thus we can to some extent at any rate picture to ourselves the process by which the wonderful order, social economy and prolonged vitality of the community has been brought about in the case of the honey-bee.

The majority of the bees are clad in rather sober colours; but some, e.g. *Sphecodes*, are black and red, others, e.g. *Nomada*, yellow and black or yellow and red. It is a curious fact, for which I can offer no explanation, that those bees which are "cuckoos" in habits are for the most part more conspicuously dressed than those which are more respectable and conscientious parents. It is not so surprising that the "cuckoo-bees" are generally, though not always, comparatively smooth and destitute of the hairy covering which is so common a feature in most bees. The presence of hairs, and especially of plumose hairs, is no doubt of great value in entangling and collecting from flowers the pollen on which all bees more or less rely: the "cuckoo" whose young is nourished at the expense of another's industry has no use for any special growth of hair and remains comparatively bald.

The most important feature in the structure of a bee, important, that is, in classifying the bees into families, genera and so forth, is the tongue and mouth

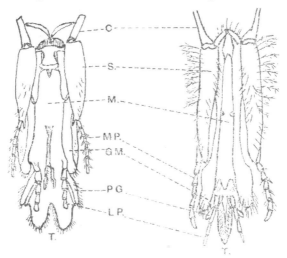

Back view of mouth parts of Back view of mouth parts of
Colletes (forked tongue)   *Andrena* (short pointed tongue)

Fig. 9.

C., cardo of maxilla. G.M., galea (lacinia of some authors) of maxilla. L.P., labial palp. M., mentum. M.P., maxillary palp. PG., paraglossa. S., stipes of maxilla. T., tongue.

apparatus (see figs. 9 and 10). There are three chief types of tongue—(i) a short forked tongue resembling that of a Fossor; (ii) a short pointed tongue, like a spearhead in shape; (iii) a long, band-like tongue whose sides are parallel for a considerable distance and then converge so as to taper off very gradually to a point. The short pointed tongue is further associated with a cylindrical shape in *all* the joints of the labial palps (*Andrenidae*); whereas in those bees (*Apidae*) which have long tongues the basal joints of the labial palps form, as it were, sheaths on either side of the tongue, the apical joints only being cylindrical. A complete description of the several portions of the mouth apparatus would lead us into deeper technicalities than the present occasion warrants: the accompanying figures (figs. 9 and 10) will, however, show their general relations.

The forked-tongued Bees.

Of the two hundred, or thereabouts, species of bees found in Britain, only seventeen possess the short, forked tongue. These are placed in the two genera *Colletes* and *Prosopis*: the former genus receives its name on account of its members lining their cells with a gluey material; while the name "Prosopis" refers to the conspicuous white markings on the faces of the male bees of that genus. With one, and that a rare, exception all the British species of *Colletes* are covered on the head and thorax with brownish hair, and the abdominal segments have dense, close bands of whitish hair encircling them. They nearly all occur in July and August, and are to

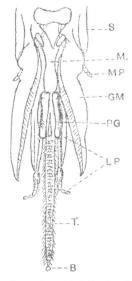

Fig. 10. Front view of mouth parts of *Bombus* (long-pointed tongue)

B., 'bouton,' or ladle-shaped tip of tongue, found only in the higher forms. G.M., galea (lacinia of some authors) of maxilla. L.P., labial palp. M., mentum. M.P., maxillary palp. PG., paraglossa. S., stipes of maxilla. T., tongue.

be found visiting flowers of heath, ragwort, chamomile and other plants; while they make their nests in the ground, generally in a rather hard sand-bank.

Frequently large colonies of them inhabit the same bank. The burrows made by them are straight, or at any rate not branched, and contain about half a dozen cells composed of the gluey, membranous material from which the genus has received its name. Hovering round, and often entering the burrows, and also associating with the *Colletes* upon the flowers there is often seen a pretty little bee, black but decorated with spots of red and white, which "plays

Fig. 11. *Colletes succincta*
(about 1½ times natural size)

the cuckoo" upon those species. The name of this parasite is *Epeolus*.

If it be desired to see excellent examples of the "plumose" hairs that are characteristic of bees, specimens from a *Colletes* should be obtained; they are most beautifully branched, and well worth microscopic examination.

The other forked-tongued bees, the Prosopids, are very different both in appearance and in habits: they are nearly all small, coal-black (excepting their faces) and almost destitute of hair. So scant is their hair, and so insignificant the brush for collecting pollen that these bees were for a time unjustly suspected of cuckoo habits. They make their nests in the stems of brambles, or of dock, or in wooden posts, or in holes in walls. When utilising plant stems they burrow down the pith, and are particularly fond of availing themselves of the bramble stems which have been cut off by hedge-trimmers; in the cut stems the pith is exposed and so the bee is saved the labour of gnawing through the wood. Examination of cut bramble stems, whether living or dead, in autumn and winter is sure to reveal holes bored by *Prosopis*, or possibly by other bees. If such stems are carefully opened the cells of *Prosopis*, recognisable by the thin skin enveloping them, are sure to be found. The cells are provisioned with honey, or with a mixture of honey and pollen obtained from flowers of bramble, mignonette, stonecrop, spurge, hawkweed and other plants. Occasionally one of the ruby-wasps may be found in "cuckoo" occupation of the Prosopid cell. Most, and perhaps all, the Prosopids emit, when handled, a peculiar, but quite agreeable scent: that of *P. signata* always reminds me of the lemon-scented verbena.

The short-pointed-tongued Bees (Andrenidae).

In point of number of species this section is by far the most numerous of our British bees, more than one hundred and twenty different kinds being known. They are grouped by entomologists into ten different genera of which *Andrena, Halictus, Nomada* and *Sphecodes* contain the greatest number of species; the two last are "cuckoo" bees, laying their eggs in the nests of the two preceding. Andrenas are the "hosts" for most of the Nomads; Halicti for the Sphecodes.

Very many of the bees of the genus *Andrena* occur early in the year, being on the wing in March, April and May: they often abound upon the blooms of the sallows, and of blackthorn. Some of these early species have a second brood which appears later in the summer, while a few are to be seen only in July and August. Of these last *A. argentata* is extremely common on the heaths of Surrey and Hampshire where it may sometimes be seen in hundreds flying over and visiting the heather bells: it is a small bee about a third of an inch long and is clothed with pale grey, almost silvery hair. All the Andrenas are hairy bees, and their abdomens are rather flat along the back: in general appearance many are like the honey-bee, but considerably smaller. One of the most noticeable of this group is a bright brown, almost red little bee, named *A. fulva*. In some places this handsome little insect is very abundant and is sure to attract attention not only by its bright colouring, but also by its habit of burrowing in garden paths and lawns, where it throws up little mounds of fine soil from its tunnel. It often occurs in large numbers quite suddenly during the month of April. These individuals which thus appear in the warm days of early spring have just hatched out from the burrows made in the previous year by their parents: the burrows are not noticeable until the bees emerge because they have been filled in with earth. The new arrivals in turn set to work making fresh burrows for their own families. Each female soon after mating digs her tunnel some six to ten or twelve inches into the earth: from the sides of the main shaft she excavates short side galleries, and fashions each of these into the form of a cell which she provisions with a mixture of honey and pollen for the benefit of the larva which will shortly issue from the egg laid in the cell. Having completed and sealed up one cell, she proceeds to the next, and so on until the burrow is nearly filled to the surface. Hence if one or two females of *A. fulva* happen to take a fancy to a particular path or piece of lawn in one season, there may issue in the following year so enormous a number of young as to excite wonder as to their origin. The white grub that emerges from the egg is full fed by midsummer and then becomes a resting pupa, similar to that of a wasp already described. As happens with several other subterranean pupae, e.g. cockchafer and stagbeetle, the imago stage is reached in late summer and early autumn; and there the perfect insect remains underground and dormant until April or May of the following year. They are not infrequently brought to light by the spade or fork in mid-winter.

The Nomad bees of whom so many are "cuckoos" upon various species of *Andrena* might easily be mistaken for small wasps, their colours being yellow and black, or yellow and red. Although so different at first sight from the usually sober-coloured Andrenas, they present both in the pupal and in the imaginal stages so many points of structural similarity to their hosts that it is thought that they are a stock of the *Andrena* family that has become decadent as the result of disregarding the plain duty of every parent. No entirely satisfactory explanation has yet been given to account for the gay uniform of the Nomads: it is possible that yellow and black is in evil repute among insectivorous creatures, and that the comparatively defenceless Nomad—her sting is a contemptible weapon—profits by the resemblance to really formidable or nauseous insects; but it is impossible to feel that this exhausts the question. A very curious difference between the Andrenas and the Nomads lies in their death attitudes: when killed by chloroform an *Andrena* always bends its abdomen downwards and forwards beneath the thorax, whereas a Nomad killed by the same method always turns its abdomen upwards above its thorax. I can offer no explanation of this peculiarity.

The *Halicti* are also burrowing bees, frequenting banks, sandy places on heaths, hard gravel paths and similar situations; and often forming large colonies. They are moderately hairy; most are dark in colour with narrow bands or spots of whitish hair on the abdomen, which is not so flattened as in the Andrenas. Some however have a bright brown and smooth abdomen, and a few are of a bronzy green colour.

Fig. 12. Last segment of abdomen of female *Halictus*, showing the ridge ("rima") characteristic of females of this genus.

Some of the smallest of our British bees are included in this genus, several species of *Halictus* being less than a quarter of an inch in length. The female *Halicti* can readily be distinguished by the possession of a very well-marked ridge (the *rima*) along the centre of the back of the fifth segment of the abdomen; the sixth segment is hardly visible, so that the "rima" appears to be on the terminal segment (see fig. 12). The sting is too feeble to penetrate the human skin. The various species are to be found in the warm months of the year from April to as late as early October; and many (perhaps all) are to be found at the beginning and again at the end of the season. They are very partial to composite flowers such as dandelion, hawkweed, etc.

Some of the species of this genus are of special interest in showing a tendency towards the social condition. Fabre's observations on *H. lineolatus* and *H. sexcinctus* establish that there is considerable collaboration between the members of a colony. A number of individuals work together in sinking a main shaft which branches out into side galleries beneath the earth, the galleries leading to various groups of cells. Each such group is the work of one female; so that a considerable number of nests are served by one and the same entrance and corridor. A sentinel is often stationed at the entrance, and there is close behind the doorway a small recess into which the sentinel can step whenever she wishes to allow a member of the establishment to pass in or out. The seasonal distribution of the sexes of *Halicti* is sure to strike any observer or collector of these insects: in the spring females only are to be found, while later in the year there is an astonishing preponderance of males. The explanation of this circumstance shows a further step towards the habit which prevails among the social bees. The females that appear in the spring have hibernated in their burrows, but not after the same fashion as the Andrenas which do not issue from their burrows at all until the spring. These female *Halicti* were hatched out in the late summer of the previous year, at the same time as their brothers and husbands who are so much in evidence at that season; after pairing—and there is evidence that pairing often takes place within the burrow and that the female (like Andrenas) does not in such cases come out into the open—the female at once retires into the burrow and there hibernates, having made no cells nor stored any provisions either for herself or her offspring. The males all die off before the winter. In the following spring the female wakes up and at once sets about the work of constructing burrows and provisioning cells with pollen and honey, and of laying eggs.

In other species, e.g. *H. morio*, a very common British form, it appears that there is only one generation in the year, but the period of emergence is very extended, lasting over several months.

Both *Halicti* and *Andrenae* are very subject to the attacks of a small parasitic insect, usually regarded as an aberrant beetle, named *Stylops*. The head of the parasite is often visible projecting between the abdominal segments of the bee; the body is lodged among the viscera of the "host" and absorbs nourishment from them. The female *Stylops* reaches maturity within the bee's body and never leaves it; the male, on the other hand, makes his way out, and flies about with much activity. The female produces large numbers of young, but we do not know with certainty how these gain access to the larvae of the bees: for it is at this stage that the host is attacked by the parasite. The general effect of "stylopisation" is the suppression of the distinctive secondary sexual characters:—affected males fail to develop their usual masculine features, and assume some of the feminine; and *vice versâ*.

The Sphecode bees are of peculiar scientific interest inasmuch as there seems no reason to doubt that they are not thorough-going "cuckoos," but are sometimes industrious in forming burrows. They are rather small, highly polished bees, with red, or red and black abdomens, and might easily be mistaken for Fossors; and their pollen-collecting apparatus is very meagre. Their usual habit is to be parasitic ("cuckoo") upon various species of *Halictus*. Structurally, and also in hibernating habits, they closely resemble *Halicti*, although so different from the majority superficially: so that it is possible that they are a degenerate offset from the *Halictus* stock. Of their bad habits there is no doubt: species of *Sphecodes* have been taken out from the brood-cells of *Halictus*; a *Sphecodes* has even been seen to attack and kill a *Halictus* and then take possession of its burrow. Moreover, certain species are always found with certain *Halicti*, and are not found in localities from which these *Halicti* are absent. On the other hand, several witnesses bear testimony to having seen *Sphecodes* making burrows for themselves. Hence the doubt arises as to whether *Sphecodes* has not yet completely fallen into parasitic habits; or, on the other hand, is on the upward path, and has only recently begun to collect and store pollen for its young. There is much need of close and patient observation of this interesting genus before a final verdict can be pronounced.

Fig. 13. *Dasypoda hirtipes*
(about twice natural size)

This chapter would be incomplete without some reference to the most beautiful not only of the Andrenids, but of all our British bees, *Dasypoda hirtipes*, as it is aptly, though somewhat pleonastically named. This is a large bee, not uncommon in sandy places, and in shape is very like an *Andrena*. The head and thorax are clothed with yellowish brown hair; the abdomen is black with conspicuous belts of white hair: the long hind legs of the female carry on the tibiae and metatarsi very long, beautifully branched bright golden bunches of hair, which give the insect a most striking appearance. This bounteous endowment enables the insect to carry enormous loads of pollen, as much as half her own weight, back to her nest. This is made in a deep burrow, from one to two feet in length in the ground. The soil is excavated by the mandibles; the front legs working very rapidly thrust the loosened sand backwards under the body of the insect; at the same time the middle pair of legs is employed to move the bee herself back towards the entrance; and on arrival there the hind legs sweep away the earth on either side. The whole of this process is rapidly repeated time after time until the desired depth is attained. About six brood chambers are made along the sides of the burrow, and each is provisioned in turn, the lowest first. To this end several large loads of pollen are brought to the cell, moistened with honey, and then moulded into a ball; another load is added as an outer shell to the ball and three short feet are fashioned so as to raise the main mass clear of the floor: this accomplished an egg is laid on top of the mass. A second chamber is then hollowed out, the material so obtained being used to close the now fully equipped first cell. The larva which hatches from the egg lies in a curve on the top of the food-mass, and devours it layer by layer so as not to disturb the spherical shape. No excrement is voided until the whole of the food-mass has been eaten (cf. Wasp larva). Having accomplished these acts the larva rests motionless, but does not become a pupa for some months. The pupal period is comparatively short, and is remarkable for the amount of movement displayed by the insect at this stage.

CHAPTER VI

LONG-POINTED-TONGUED BEES (APIDAE)

THIS section includes the greater number of our British genera; but, since the genera contain but few species, the total number of kinds occurring in our islands is less than fifty. The most noticeable among them are the Leaf-cutter bees (*Megachile*), the Mason bees (*Osmia*), the Wool-carder bee (*Anthidium*), the Long-horned bee (*Eucera*), the Humble- (or Bumble-) bees (*Bombus*), and the Honey-bee (*Apis*); the two last named being "social" in their habits. In addition to these there are several less well-known genera of industrial habits, and about half a dozen which are "cuckoos," including *Epeolus* already mentioned as parasitic upon species of *Colletes*.

As already stated, all the *Apidae* have long tongues with sides parallel for a considerable distance and then gradually converging to a point, and the basal joints of the labial palps form a sheathing investment to the base of the tongue, the apical joints only being cylindrical. The position of the pollen-gathering apparatus varies greatly in the different genera; in some, e.g. *Apis, Bombus, Anthophora*, the hairs for this purpose are carried on some of the joints of the legs; while in others, e.g. *Osmia, Megachile, Chelostoma, Anthidium*, the ventral surface of the abdomen has a densely hairy coating, which is often brightly coloured.

A bee belonging to this group is probably known by sight to most country-dwellers, though it is very likely mistaken for a small and unusually active humble-bee. We refer to *Anthophora pilipes*, which is an invariable herald of the advent of spring, and may be seen on any bright sunny morning in March or April (I have seen one on the wing in January) dashing from crocus to crocus flower, or hovering over the early blooms of *Arabis, Aubrietia* and other plants. The female is black and hairy with reddish-yellow legs, and is about half an inch long: she diligently visits the flowers in search of nectar and pollen. The male is, when fresh, bright brown, but his colour soon fades to greyish-brown: he has extraordinarily long grey and black hairs projecting from the tarsal joints of the middle pair of legs: he seldom settles, but flies about in attendance on a female, often chasing her, or appearing to join in a wild game of "catch-as-catch-can" with one or two other males. These bees make their nests in firm banks of sand or clay, if not too wet; their burrows do not extend very deep and contain one or more cells whose outer wall is made very hard by a cement of sand or clay applied by the female bee after she has completed the commissariat arrangements. Numbers of *Anthophora* often live in the same bank; and on a warm April morning the scene at such a spot is most lively: females are to be seen entering or leaving their burrows, intent on their business, or possibly engaged in a headlong flirtation with two or three males to and fro in front of the favoured bank, while dozens of males keep up a loud humming as they dash through the air.

If one of these colonies of *Anthophora pilipes* be found, another bee is also certain to be present playing "cuckoo" upon the legitimate tenants of the burrows. This parasite is named *Melecta armata*: it is nearly as long as its "host," but very different in appearance, being nearly black with white spots on the sides of the abdomen, much less hairy, and far more sharply pointed at its hinder end. Anatomically, however, in the arrangement of the mouth parts and of the male sex-organs it is closely allied to its host. I have never seen *Anthophora* offer any resistance to *Melecta*, nor indeed take any notice of its disreputable cousin. We can hardly imagine that the bee knows that the egg she lays (if indeed she knows that she lays an egg) in her cell is one day destined to produce a bee like herself; still less can it be thought that she appreciates the fact that the presence of *Melecta* involves disaster for at least one member of her family.

The bees of the genus *Osmia* are singularly versatile in their habits, and adapt themselves with great ingenuity to many very different nesting sites. Many instances are recorded in which individuals of the same species select an extraordinary variety of places for their nests, thus showing that they are following no blind instinct, but have some power of choosing for themselves. For example, *Osmia rufa* (the "Mason Bee"), a little bee about half an inch long and covered with yellowish-red hairs, will sometimes nest in the crevices of an old brick wall; at other times it will burrow in the ground; at others it will employ old snail-shells. In the Natural History Museum at South Kensington there is exhibited a flute within which *O. rufa* had constructed fourteen cells; while in the Charterhouse School Museum there is an outhouse lock which was entirely filled up with the mud cells of this species during the course of the summer holidays: I have also found

VI] LONG-POINTED-TONGUED BEES 77

this species burrowing in partly decayed wooden posts. The female *O. rufa* is readily recognisable by the possession of stout horn-like outgrowths on the face. Another less common species, *O. leucomelana*, usually tunnels its way into the pith of bramble stems, but is recorded by the late Mr Ed. Saunders as nesting in the side of a sandy road. *O. bicolor*, again, nests either in banks or in empty snail-shells. Mr V. R. Perkins observed that in some instances after the snail-shells had been filled with cells, the little bees covered them up with short pieces of "bents" so as to form a small hillock two or three inches in height and about six inches round the bottom. Mr F. Smith, in his *Catalogue of British Hymenoptera in the British Museum*, tells of an *Osmia* which, having selected a large snail-shell for her nest, on reaching the wider whorls of the shell departed from the usual custom of building the cells in single file, and placed two cells side by side across the cavity. Many other instances might be adduced of species of *Osmia* adapting their behaviour to meet special circumstances. They are undoubtedly among the most intelligent of all the solitary bees.

The *Osmiae* are all rather short, stout bees—perhaps the word "cobby" best describes their general build: they possess very long tongues, and all their mouth parts are decidedly elongate. They are thus able to reach to the bottom of such deep flowers as ground ivy and thistle in search of nectar. The pollen-collecting apparatus of the females is in the form of a dense brush of hairs, often very brightly coloured, upon the under side of the abdominal segments.

The rare "cuckoo" bees of the genus *Stelis* are parasitic upon some species of *Osmia*.

We have in this country several other genera of bees in which the pollen brush is situated as in *Osmia*. Of these, *Chelostoma* contains two species, both of which are so small that they are not very likely to attract attention, though one of them, *Ch. campanularum*, may often be seen in large numbers hovering round and entering the flowers of the harebell and various species of "Canterbury bell" in gardens during June and July: both species frequently are to be found in the evening asleep in the recesses of the bell-shaped blossoms. The true leafcutter bees of the genus *Megachile* are, however, large and conspicuous insects, and often force themselves upon our notice. The habit of using portions of leaves for the construction of the nest is not entirely confined to *Megachile*; for, as might perhaps be expected, one species of *Osmia*, viz. *fulviventris*, is stated on the authority of that excellent observer, Mr R. C. L. Perkins, to cut out for this purpose portions of slightly withered, yellowish leaves, though not with the deftness of a *Megachile*, for a jagged,

vi] LONG-POINTED-TONGUED BEES 79

irregular piece is removed, comparing badly with the neat work of the true leaf-cutters.

There are eight British species of *Megachile*: they are all decidedly hairy bees (about the same size as

Fig. 14. Portion of a rose-leaf which has been attacked by a leaf-cutter bee (*Megachile*)

the honey-bee, but of stouter build), with large wide heads and powerful flattened mandibles; their labrum (upper lip) is long and square-cut, and capable of complete inflection beneath the head so as to be concealed by the mandibles, and the tip of the abdomen is very blunt. The males of some of the species have curious flat expansions on the tarsal joints of the front legs. For purposes of nesting burrows are made in old posts, branches or stumps

Fig. 15. Three cells of leaf-cutter bee (*Megachile*) in a piece of rotten wood

of trees, grassy or sandy banks; some species adhering closely to one kind of site, others adopting sometimes one, sometimes another. They all line their burrows with neatly cut pieces of leaves, e.g. rose, privet, or of the petals of flowers, e.g. geraniums, bird's foot trefoil, etc. When the nest is made in wood, the female insect bores the burrow herself, though she is ready enough to avail herself of one that has been tenanted the previous year: and similarly when the nest is subterranean, the bee will make use of a disused worm-burrow or will, if necessary, drive a new tunnel for herself. So far as I have observed, the burrows whether in earth or wood are never straight, and always have an upward inclination; or, if winding and at first slightly downwards, then there is a sharp upward bend near the entrance. This arrangement prevents rain from soaking the burrow in its brood portion. The cells constructed by these bees resemble little thimbles in shape, and are wonderful works of art. They are composed entirely of pieces of leaves or of petals slightly glued together by a waxy excretion produced by the bee herself. Each piece appears to be cut with a special view to the particular place which it is destined to occupy in the completed cell. The side pieces are of a rounded oblong shape, and their size is gradually decreased both in width and length as layer after layer is added within those which actually touch the walls of the burrow. In a specimen taken from the nest of *M. circumcincta* I find twelve such side pieces; the dimensions of the innermost are 0·60 inch × 0·32 inch; of the sixth from the inside, 0·65 inch × 0·35 inch; of the outermost, 0·80 inch × 0·40 inch. All these pieces include a portion of the naturally serrated edge of the rose-leaf, a provision which appears to make the overlapping junction of piece to piece more secure. The two ends of the cells are composed of circular, saucer-shaped pieces; and these seldom, if ever, include any of the marginal serrations: they lie inside the side pieces, and it is thus important that their circumferences shall be as smooth and regular as possible, in order that no gaps may be left to afford entry to small thieves. The bottom pieces are placed with the hollow of the saucer inwards, while those at the top present their convex surface to the space within the cell, thus leaving the upwardly directed concavity ready to receive, or rather to help in forming the bottom of the next cell to be constructed. In the specimen before me there are only two pieces at the bottom, but six at the top: they are all so perfectly circular that with an instrument capable of measuring accurately to $\frac{1}{100}$th of an inch, I can hardly detect any differences in the diameters of any one: the diameter of the upper bottom one (i.e. that nearest to the cell-cavity, and therefore the smallest), is 0·30 inch; that of the innermost lid is 0·33 inch; and that of the topmost lid 0·35 inch. It is most interesting to watch one of these Megachiles at work upon a rose-leaf: having alighted upon a leaflet, she holds on with her legs to the edge of the piece which she wants to cut off and then, by the combined action of the mandibles and chisel-like upper lip, rapidly makes the necessary incisions: directly the piece is free, by means of her wings she saves herself from falling; takes (in the case of a side-piece) a firm grip of the middle of one of the short sides with her mandibles, thus making a slight crease along the centre and folding down the edges of the long sides; with her six clawed feet, three on each side of her burden, she holds the piece along these edges securely under her body, and flies off to her nest, looking like some large green insect as she speeds through the air. The cells are stored with a mixture of pollen and honey of such quality and quantity that in some districts children dig out those of *M. maritima*, and devour their contents. The sting of this species, and indeed of all in this genus, is incapable of piercing the human epidermis. In districts where leaf-cutters are plentiful, and they are fairly common in most parts of England, it is a frequent occurrence to find rose trees whose leaves have been freely attacked: the work of a *Megachile* is always recognisable by its neatness and regularity. I have alluded above to the remarkable dilations of the tarsal joints of the anterior legs of some males: this peculiarity is very noticeable in *M. maritima*. The use of these extraordinary outgrowths is not fully known: I have observed that *M. maritima* is very fond of "washing his face" with them, but this is not an adequate explanation of their presence.

Most of the leaf-cutters suffer from the attention

of various species of "cuckoo" bees belonging to the genus *Coelioxys*. In structure, as is so frequently the case, these "cuckoos" closely resemble their hosts, but their superficial appearance is very different. In size they are somewhat smaller; in colour they are black with conspicuous white bands; while the abdomen is almost destitute of hairs, and in the females is very finely pointed: the abdomen of the males ends in a series of spines. In the case of solitary bees, like *Megachile*, which do not form colonies after the fashion of *Anthophora*, the question inevitably occurs, "How do the 'cuckoos' find the nest of the bee that is to foster their young?" A lucky chance enabled me to throw some light on this point. One day towards the end of June, 1906, I happened to see a *Coelioxys quadridentata* enter the burrow of a *Megachile circumcincta*. I dug the nest out of the burrow and in so doing scattered the sand over an area of several square inches, completely effacing all trace of a burrow. I awaited the return of the *Megachile* in order to identify the species, and was astonished in the course of the next ten minutes to see and capture two more specimens of *Coelioxys* which came up against a fresh breeze that was blowing at the time, and alighted on the disturbed soil. It is evident that these "cuckoos" must have been attracted to the spot by the scent of the excavated nest. I may further mention that *Coelioxys* was not abundant on that occasion, for

I saw no other specimens during several hours spent on the same heath that day.

Of the remaining genus of "hairy-bellied" bees, *Anthidium*, we have only one species, *A. manicatum*, the wool-carder bee, in this country. It is a striking looking insect, about three-quarters of an inch long (the male being larger than the female), and is black with conspicuous yellow spots, so as to appear rather wasp-like. It is, however, a stoutly built bee, with blunt hinder extremity, that of the male being armed with five strong "teeth." This bee is on the wing in June, July and August: it uses many different sites for nesting, e.g. deserted burrows made by other insects, door-locks, empty snail-shells, and so on. Smith states that it has never been known to make a burrow of its own; this may be so, but inasmuch as I found one nesting under the slates of an outhouse which was in process of being roofed, it is evident that it does select new nesting sites sometimes. The nest is like a ball of white wool with waxy cells within it. This "wool" is obtained from hairy plants, especially the purple dead-nettle; but I have known the bee to avail itself of a boy's "sweater" left hanging in a school pavilion. It is of this bee that Gilbert White, in his *History of Selborne*, writes: "There is a sort of wild bee frequenting the garden-campion for the sake of its tomentum, which probably it turns to some purpose in the business of nidification. It is very pleasant to see with what address it strips off the pubes, running from the top to the bottom of the branch, and shaving it bare with the dexterity of a hoop-shaver. When it has got a bundle almost as large as itself it flies away, holding it secure between its chin and its fore-legs."

Before proceeding to the social-bees, a passing mention must be made of two very striking solitary

Fig. 16. *Eucera longicornis*
(about 1½ times natural size)

species, viz. *Eucera longicornis* and *Saropoda bimaculata*. The former of these is a large and handsome brown bee, about three-quarters of an inch in length, that in some localities is abundant from the months of May till August. The female of this species has antennae of ordinary dimensions; but in the male these organs are nearly as long as the whole body, rendering the insect quite unmistakeable. The nest is in the earth at the enlarged end of a burrow about six inches long. *Saropoda* is nearly allied to *Anthophora* not only in structure but also in general appearance and in its swift dashing flight, but is remarkable on account of the lovely blue-green colour of its eyes, and the peculiarly shrill note of its "hum"; this last is so characteristic that it is quite easy to recognise the presence of the bee without seeing it. The bee appears in July and August, and burrows in the ground, large numbers often associating together to form a colony.

CHAPTER VII

THE SOCIAL-BEES

THE humble-bees, of which there are more than a dozen British species, are perhaps the most generally familiar of all the Hymenoptera; and probably most people have noticed that there are several different types of colouring and different sizes of body among them. Speaking broadly, the largest individuals are perfectly fertile females ("queens"), the smallest are sterile females ("workers"), while those of intermediate size may be either males ("drones") or females whose fertility is but partial. None of the smaller females exhibit any structural distinction from the "queens"; it is only in size that they differ: and in this respect it is to be observed that the humble-bees resemble the social-wasps more nearly than they do the honey-bee. Colour, unfortunately, is no sure guide in distinguishing one species from another; for the colours of the hairs vary so greatly in different individuals of some species that structural features alone are to be trusted in diagnosis; and these last are often obscure and difficult to appreciate. However, the large black humble-bee with a red "tail" will generally prove to be a "queen," *Bombus lapidarius*; that with a yellow band across the thorax, another near the front of the abdomen, and with a greyish-tawny "tail," *B. terrestris*; that of a tawny yellow colour all over, *B. venustus* Smith (*muscorum* Kirby), while that with a tawny thorax and black, or black and tawny abdomen, *B. agrorum* Fab (*muscorum* Smith). Meanwhile, it should be stated that if the insect has smoky black wings, then it is not an industrious *Bombus*, but a lazy "cuckoo," *Psithyrus*, of which there are several species closely resembling the more virtuous *Bombi* on whom they foist their young.

According to their nesting habits the *Bombi* are divisible into two groups: (i) those which nest beneath and (ii) those which nest on the surface of the ground —"carder bees." *B. lapidarius* and *terrestris* are examples of the former; *B. venustus* and *agrorum* of the latter group. The underground species form far larger communities and are much more pugnacious than the surface-dwellers: there is no truth in the popular belief that humble-bees cannot sting. The history of a single community is not very different in the two groups, and bears a decided resemblance to that of the social-wasps; for at the end of the season the fertile, impregnated young females hibernate, and the rest of the population perishes. The female, awakened by the first warm days of early spring, searches along hedgerows and banks for a suitable spot either on or in the ground where she may form her nest. Mouse-holes, mole-runs, birds' nests occasionally, and perhaps chance hollows in the ground serve their purpose; though it is doubtful if *Bombi* ever start the nest *ab initio*, but do not rather take possession of a site that has previously been tenanted by some other animal. Our knowledge of the habits of these bees is chiefly due to the observations of Huber, Schmiedeknecht, Hoffer, Sladen and a few others. The female, having selected her nesting place, collects a small bundle of moss; beneath this she places a cell formed of wax on the outside and a lining of pollen saturated with honey: several eggs are then laid in the cell, and it is then closed. After a few days' rest, a second cell is placed beside the first, and perhaps a third, all being fastened together by a brownish mixture of pollen and wax, and all containing several eggs. Very soon the larvae hatched out in the first cell devour the slight store of food which it contains, and require to be fed by their parent. For this purpose a small hole is made in the lid of the cell, and through this the mother discharges food from her mouth for the benefit of her offspring. As the larvae increase in size they distend the cell in a number of irregularly bulging pockets, so that it now appears a very misshapen affair. When the larvae are full-fed they spin a fine silken cocoon and pupate within the cell, but not simultaneously in any one cell although the eggs from which they arose were all deposited at approximately the same time; neither do all the perfect bees issue from a given cell at the same date, even if all are of the same sex. These first cells all give rise to small females ("workers"), whose escape from the cell is made easier by the parent bee removing the wax with which the cocoons are covered. As with the social-wasps so too with the humble-bees, as soon as the foundress of the society has reared a batch of "workers," she seldom if ever quits the nest, but henceforth confines herself to the work of egg-laying: in some cases she actually loses the power of flight. With the increasing strength of the working staff of the establishment the number of cells is quickly increased; but the architectural skill of the *Bombi* is of a very low order when compared with that of either social-wasps or honey-bees, for the building proceeds upon no definite plan or system. The cells which have once been used are not employed a second time for nursery purposes, but new cells are placed upon them in haphazard fashion, so that an old nest presents a very irregular, knobby appearance, which is enhanced by the varying sizes of the cells themselves. In the smallest cells "workers" are reared, in the largest fertile females, and in those of intermediate size, drones. It appears also that a nourishing lining of pollen and honey is provided only to the earliest cells: the larvae of those of later date are fed entirely by the workers from day to day. To safeguard the rising generation, and also the full-grown inhabitants of the nest, against famine during bad weather stores of pollen and of honey are laid up within the nest. The old cells and the empty cocoons from which bees have emerged are employed as tubs for this purpose, and special waxen honey pots are constructed and left permanently open, *pro bono publico*—a fact of which small boys at times avail themselves. The small females which are produced during the middle of the season frequently, perhaps always, become fertile, and together with some of the "workers" take some part in egg-production, and thus add to the numerical strength of the colony. We have very little knowledge of the causes which determine for or against the fertility of these small females, neither is the sex of their offspring known with certainty; but there is no doubt that in the event of the premature decease of the foundress "queen," the colony can still continue, thanks to the fertility of these her former servants. It is only towards the end of the summer that the drones make their appearance, ready to mate with the young large females which are alone destined to pass through the winter. The nests are at all times liable to the attacks of mice, earwigs, beetles, and many other insects, and are quickly brought to nought when autumn sets in.

Despite the fact that the *Bombi* are but inferior architects, they are nevertheless a most industrious people. The working day begins with them at three or four o'clock in the morning, and is continued till dusk or even into the night, if the weather be warm and the moonlight good. It was stated by Godart, some two centuries ago, that some species appoint a trumpeter-bee to rouse the rest of the community; this statement was confirmed in the latter part of the nineteenth century by Hoffer, who observed the performance in a nest kept in his laboratory, and noted that when the trumpeter-bee was taken away its place was filled by a substitute the next morning. The doubt which was for so long entertained with regard to this trumpeter is perhaps due to the fact that most human beings receive their summons to leave their beds several hours after the humble-bees have been "called."

VII] THE SOCIAL-BEES 93

In point of numbers a *Bombus* community falls far below other species of social Hymenoptera. A strong subterranean nest may have a population of about 300 or 400 individuals; while that of a surface-builder seldom contains more than half that number, and frequently much fewer. Smith gives statistics of a nest of *B. muscorum*, which yielded 25 females, 36 males and 59 workers.

The number present is largely determined by the amount of attention received from one or other of the five species of *Psithyrus* which play "cuckoo" upon the humble-bees, especially upon the subterranean species. As a rule the *Psithyrus* both in appearance and in structure closely resembles the *Bombus* which it patronises, but is of larger dimensions; for example, *Ps. rupestris* invades the nests of *B. lapidarius*, and like its host is a large black bee with a red tail; the "cuckoos" are, however, entirely destitute of all apparatus for the collecting of pollen, and only occur in the form of males and large females: there are no "workers." The *Psithyrus* is admitted to the nest of the *Bombus* without any demur and actually goes so far as to build cells, which are recognisable by their greater size, for the reception of her eggs. Hence we may presume that the "cuckoo" habit has been but recently acquired, and that the call of duty has not quite ceased to make itself heard. Beyond this, however, *Psithyrus* does not go. She leaves the work of nourishing her young to the *Bombus* workers; she does not bestir herself till the morning is far advanced, and returns, empty-handed, early in the evening; she may even, in the middle and later part of the season, assume to herself the privileges of the lawful "queen," and stay at home all day moving about within the nest, as is the way with the "queen," but, instead of laying eggs, eating up the stores of pollen and honey intended for others. It is this wholesale robbery of stores that causes the marked diminution in the *Bombus* population.

At the onset of autumn the young female *Psithyri*, having mated, hibernate in various snug retreats, ready, if no Nemesis befalls them, to continue the evil work of their mother in the following summer.

From an economic, agricultural point of view the humble-bees are of great value to the farmer and gardener. The flower-loving habits of all bees render them most important agents in the pollination of flowers and in the consequent "setting" of the fruit: it is indeed recognised that hives of honey-bees are an almost indispensable adjunct to a successful fruit farm. The various wild bees play their part in this matter; but the humble-bees being strong, powerfully built insects, and provided with very long tongues, are able to visit, with satisfaction to themselves, many of the tightly closed papilionaceous flowers which demand for their opening more strength than is possessed by any other bees, and to reach the nectary at the bottom of flower-tubes too deep for the shorter tongued bees. A notorious instance of the agricultural importance of humble-bees was afforded when first the red clover was introduced into New Zealand: there being no indigenous bees capable of pollinating the blossoms, it was found impossible to raise any seed. But now that the humble-bees imported from this country have become naturalised in New Zealand, the seeds are "set" with complete success. It may not be out of place here to remark that since field-mice are most persistent and destructive enemies of humble-bees, it is most unwise of farmers to attempt to exterminate such creatures as weasels, owls and the smaller hawks which feed largely upon these little rodents, who are also destructive to crops in other ways.

The structure, life-history and habits of the honey-bee, *Apis mellifica*, which stands at the head of the anthophilous Hymenoptera, has been so fully dealt with by many authors that we do not propose here to give more than a very brief outline of this most important species.

A flourishing hive of honey-bees will contain, in summer time, a fertile female (the "queen" bee), an enormous host of sterile females (the "workers"), and a number of males ("drones"). The material of which the nest, i.e. the combs, are composed is wax which is produced by glands on the ventral side of the abdominal segments of the workers. This wax passes through some peculiar membranes on the under side, and makes its appearance in the form of thin plates projecting between the segments. From this situation the plates are removed by means of a nipping apparatus with which the hind legs of the workers are provided; they are then conveyed to the mandibles, and by these moulded and worked up to form the hexagonal cells in which food, honey or pollen, is stored and the larvae reared. To produce wax a worker must first consume large quantities of honey. It is stated that every pound of wax produced represents a consumption of about 15 lbs. of honey. The combs are all placed vertically, and the cells composing them are in horizontal rows one above the other and in double series, back to back: the outer ends of the cells are at a slightly higher level than the contiguous bases, and thus the honey has no tendency to flow out. When a cell is completely filled, its opening is sealed over with wax.

The life-history of any one "stock" of bees is a very different matter from that of any of the other social Hymenoptera, for the colony is not started by the unaided efforts of a "queen," but by a very large number of workers who have, accompanied by the "queen," left some already existing and probably overcrowded hive as a "swarm." Under normal circumstances, the "queen" which emigrates with the swarm is the reigning and therefore impregnated "queen" of the hive from which the excited throng issues. The phenomenon usually occurs on a warm sultry day in May or June; the bees pour forth from the hive in a living torrent and quickly fly up into the air where they wheel around in such numbers as to cause a small cloud, all the while buzzing in a tone that is quite characteristic of swarming to the ears of the bee-keeper. After a longer or shorter period they begin to gather in a dense cluster on a neighbouring branch or twig upon which their "queen" has already settled: "settling" may be accelerated by spraying the flying bees with water discharged from a fine syringe, or by throwing handfuls of dust up among them. Previously to swarming the bees have consumed large quantities of honey, so that each of the workers is more or less gorged, and is capable of existing thus without further food for a considerable time: they are also very loth to use their stings when swarming, and, so far as my own experience extends, the sting if used is comparatively innocuous: hence a swarm, once settled, may hang on the branch for an astonishingly long time, especially if the weather change for the worse. I have known a swarm remain suspended for four days. If left undisturbed the bees after a few hours' suspension fly off to some suitable retreat within a hollow tree, under a roof, or perhaps even to an empty hive: there is evidence that in some cases their future home has already been selected before the swarm breaks up—possibly before it has left the hive. Under the usual conditions of domestication, however, the swarm is "taken" by the bee-keeper, who may shake the bees off into a straw "skep" or other receptacle which is then placed inverted on a sheet laid on the ground in some shady spot, one edge of the skep being slightly raised to allow any disturbed bees to crawl in and join their fellows; or perhaps the branch to which the bees are clinging may be cut off and thus the swarm conveyed away bodily to the hive intended for it. The size and weight of a swarm vary very much: a poor swarm may be no bigger than the fist; on the other hand I have taken a swarm which, when held out, extended from my shoulder to my ankle, and required the use of both hands to hold it up. An average swarm is about the size of a football, and weighs about 4 lbs. The swarm is usually housed in its permanent quarters during the evening of the same day; a sheet is spread in front of and leading up to the entrance of the new hive, and on this the bees are thrown out from the "skep." They very soon walk up into the proffered shelter. It is astonishing to notice how quickly they become aware that their "queen" has gone into the hive: on one occasion a swarm that I was endeavouring to "hive" in the above fashion behaved in a refractory manner and did not go in as readily as was to be expected: search among the crawling mass brought to light the "queen" who had been held prisoner in a dense ball of clinging bees: so soon as I had put her at the entrance to the hive she walked in, and immediately with one consent, like a regiment of soldiers turning to "right-about," the whole concourse turned their faces to the hive and trooped in after her.

Once in their new quarters the bees immediately begin to secrete wax and build combs, suspending them from the top, or from bars provided for the purpose; not infrequently they start comb-building while waiting in their temporary "skep." The "queen" lays an egg in each cell, and as these quickly hatch, the workers undertake the heavy work of feeding the resulting larvae. For this purpose they themselves devour honey and pollen which are worked up into a nutritious "pap" ("chyle food") by their digestive organs, and then regurgitated and administered to the larvae. The "pap" is poured into the cells, so that the larva is actually bathed in it and in all probability absorbs nourishment through the delicate skin as well as by the mouth. When, after several moultings of the skin, it becomes full-fed the larva is imprisoned by a porous lid of pollen and wax. It then spins a cocoon within which it pupates, and from which in due course it issues as a young worker bee. For a while the young bee remains within the hive and acts as a feeding-nurse to the larvae; later on she goes out from the hive in quest of nectar and pollen which she brings back to store each in separate cells, or of "propolis"—a gummy, resinous substance collected from the buds and bark of trees—which is used by bees as a cement or to plug up undesired chinks and crevices. The development from egg to perfect insect occupies about three weeks. When another "queen" is to be produced, the workers form one or more "royal" cells: these are generally, but not always, placed at the lower edge of a comb, and are quite different in shape from the ordinary cells, being more or less flask-shaped structures of considerable size, and having their entrances facing downwards: moreover the walls of these "royal" cells are of much tougher material than those of the rest of the comb. An ordinary egg, in no way different from those which yield "workers," is laid in the "royal" cell; but the larva which emerges from it is tended with especial care and is fed throughout its life on a "royal jelly." This jelly ("pap") is also supplied to worker-larvae for the first three days after they are hatched, but they are subsequently weaned, and thenceforth fed upon honey, pollen and water: whereas the queen-larva is supplied with "royal jelly" until it is full-fed. The result of this treatment is that the development of a "queen" takes but sixteen days from egg to perfect insect, and that the reproductive organs of the insect are functional. It is not known with certainty whether the "pap" or "jelly" supplied at the first to workers is of exactly the same chemical character as that supplied to a "queen."

Drones are raised in cells of the same shape, but of rather larger size than those employed for workers. There is no doubt that the majority of drones issue from eggs from which the "queen" has withheld the fertilising male element stored within her body; but it has not been proved that they cannot also be produced from fertilised eggs. In cases of hybridisation the drones, as a rule, have all the characteristics of the race to which the "queen" belongs: they appear uninfluenced by the drone with which she was mated. Drone-larvae are said to receive rather more "pap" than do worker-larvae; and their development occupies three or four days longer. The drones all die or are killed by the workers before the winter begins. The lower portions of the combs are, as a rule, alone employed for brood purposes; the cells immediately above and to the sides of the brood contain pollen, the upper being reserved for storage of honey. It is these great stores of food which enable a "stock" of bees, alone of the insects with which we are here concerned, to continue through the winter. In a healthy hive thousands of workers, as well as their "queen," live in a more or less torpid condition through the severe weather; on milder days they will rouse themselves, feed on the accumulated stores, and even take short flights during which they rid themselves of their evacuations.

The hive from which a swarm has gone forth contains the residue of the stock of workers, and, as a rule, several "sealed" queen-cells containing queen-larvae or queen-pupae. For a time therefore this hive is without a functional queen. It is not until the eighth day after swarming that the most advanced of these young queens hatches out. Usually, but not invariably, she, with the assistance of the workers, slays her younger rivals who have not yet emerged from their cocoons. The successful queen is as yet an unmated virgin. So soon as she is strong enough she leaves the hive on her "nuptial flight" and pairs in mid air with a drone whose life is sacrificed in the act of mating. The now fertilised "queen," if no accident befall her, returns to the hive and assumes her duties of egg-laying. One act of mating is sufficient, the store of semen lasting for several years. Among domesticated bees, second and third swarms— "casts" as they are called—may be sent out from the same hive; but it is doubtful if this ever occurs in the wild condition.

It will be appropriate here to call attention to the difference between *honey* stored by bees and *nectar* produced by flowers: the two terms are often used as synonymous, in spite of the fact that anyone who has ever tasted a drop of nectar from any flower must have noticed that the limpid drop was very different from the honey of a honey-comb. Both liquids, it is true, are sweet to the taste in consequence of the sugar which they contain; but whereas nectar is a thin fluid with a high percentage of water, and has generally a flavour which distinctly suggests the flower whence it came, honey is much thicker, with far less water, and does not possess the odour or flavour of any particular blossoms—though honey derived from certain kinds of flowers is recognisable by an expert. These differences between the raw nectar and the finished product, honey, are brought about partly within and partly outside the bodies of the workers. The nectar is sucked up by the long tongue of the bee, and is received into a portion of its digestive apparatus known as the honey sac. It is probable that a portion of the water is here removed from it, and that a slight chemical change is effected also. On the return of the bee to the hive the now denser liquid is discharged from the mouth into the cells, and at the same time the secretions of certain glands in the head of the bee are mixed with it. Formic acid has been demonstrated to be present among these secretions, and this probably serves as an antiseptic and prevents decomposition of the honey. The honey, however, is not yet "ripe," and has not yet reached its final consistency; it is still too limpid. To promote further evaporation of water some of the workers undertake the task of improving the ventilation of the hive. A number of them marshal themselves in lines near the entrance, and by rapid vibration of their wings drive currents of air into and out of the hive and over the surfaces of the combs. At such times a strong current of warm air may easily be felt coming out of the hive, if the hand be quietly brought close to the entrance. This process is continued all night to a greater or less extent, and is the cause of the buzzing that may be heard inside any healthy hive long after dark in a summer night. The matured honey contains about 12 per cent. less water than the raw nectar, and is free from the volatile oils which give to nectar its characteristic scent or flavour.

In accordance with the great divergence of duties there are also very marked differences of structure between the fertile queen and her sterile industrious daughters. As already stated, the workers are endowed with the power of secreting wax; the queen also possesses wax glands and the membranes through which wax might be expressed, but both these are in a rather reduced condition and, so far as is known, are never actually employed. But it is in connection with the gathering of pollen that the structure of worker and queen chiefly differ: the hind legs of the worker are provided with a number of contrivances all directed to this one purpose, whereas those of the queen are destitute of any special appliances. The hairs that cover the whole surface of the bee's body

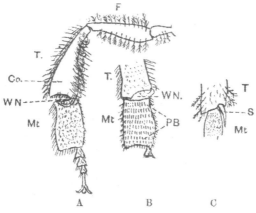

Fig. 17.

A, outer aspect of hind leg. B, inner aspect of metatarsus and lower part of tibia of hind leg. C, junction of tibia and metatarsus of middle leg of Honey-bee, showing spur.

Co., corbiculum (pollen basket). F., femur. Mt., metatarsus. P.B., pollen brush. S., spur. T., tibia. W.N., wax-nipper formed by edges of tibia and metatarsus.

are all more or less useful in enabling it to collect pollen; but those on the under side of the abdomen get most freely dusted. From this position the pollen grains are gathered by special brushes of hairs situated upon the inner face of the first tarsal joint (metatarsus) of the hind leg (*vide* fig. 17). These brushes consist of nine transverse rows of stiff, closely set hairs. When the brushes are fully laden with pollen, the hind legs are crossed while the insect is in flight (usually as it hovers in front of a flower), and the pollen is combed out by certain spine-like hairs that fringe the posterior margin of the tibia. The outer surface of this section of the hind leg is somewhat hollowed and flattened, and its anterior edge is furnished with rows of long curved hairs, so that the outer face of the tibia forms, as it were, a basket (the corbiculum), into which the pollen grains are scraped. The middle pair of legs is then employed to press the pollen firmly together in the basket, so that it may be conveyed safely to the hive. On arrival at the combs the bee pushes its hind legs into a cell near the brood, and with a spur situated on the apex of the tibia of the middle leg prises the pellet of pollen out of the basket, and lets it fall into the cell. Other bees then pack it down with their mandibles. The hind legs of the "queen" bear none of these modifications and are much less hairy; so much so that the reddish-brown colour of the actual leg is plainly visible, and affords an easy means of detecting the "queen" when she is surrounded by a throng of workers. The queen differs further in possessing a much longer abdomen; she seems in fact to trail her abdomen along with some difficulty: this feature is due to the presence of the well-developed ovaries from which her amazing multitude of eggs are produced year after year. The drone is altogether a larger insect than either queen or worker, and is of much stouter build; his eyes are very large and meet one another on the top of the head; the hair on the thorax is very dense and short, and that on the abdomen much shorter than in the workers; the hind leg is of a different shape, and has no wax-nipper; the antennae are longer; and the note of his "hum" is both louder and deeper. His sole function is to fertilise a queen.

This chapter may be closed by directing attention to the remarkable fact that in the honey-bee there are handed on to generation after generation of workers features both of structure and of habit which are not present in either of their parents. It is conceivable that in the past the fertile "queens" may have possessed pollen brushes, corbicula and all the other modifications now characteristic of the worker, but they assuredly did not possess all those instincts of the present day worker that are associated with the existence and welfare of each commonwealth of bees. These instincts can only have been evolved within the limits of the workers themselves; and yet workers are sterile and incapable of handing on

108   BEES AND WASPS   [CH.

favourable variations to offspring, for they produce none. As a study in evolution and heredity this question is full of interest.

CHAPTER VIII

SOME STRUCTURAL FEATURES: THE STING AND THE "TONGUE"

IT will not be out of place to include a description of the weapon by which the bees, wasps and ants are distinguished from other insects, and of the "tongue" which plays so important a part in the economy and on whose characters the primary classification of the bees is based.

Strictly speaking, the sting belongs to the female reproductive system, and corresponds to an ovipositor: but unlike the ovipositor of many of the lower Hymenoptera (saw-flies, etc.), it does not project from the body, but is withdrawn within it; it has moreover lost all direct connexion with oviposition, the eggs escaping at its base instead of passing through it. If a living worker or "queen" wasp be held in a pair of forceps, the insect will be seen repeatedly to dart out from the apex of the abdomen a sharply pointed structure which is usually mistaken for the sting. This organ, though sharp and strong enough to penetrate the skin is not, however, the true sting by which

VIII] SOME STRUCTURAL FEATURES 109

irritant poison is injected into the enemy, but merely the protecting sheath, known as the "director," which

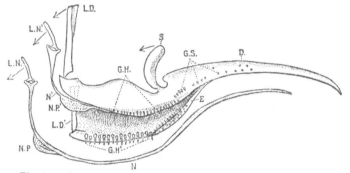

Fig. 18. Ventro-lateral view of sting of *Vespa germanica* ♀ : the needle (N.) of the far (right) side has been withdrawn and is shown below the other structures

D., director. E., ventral inturned edges of director. G.H., G.H'., guiding hairs of left and right sides. G.S., guiding studs on inner face of D. L.D., L.D'., levers of left and right sides for extrusion of D. L.N'., L.N., levers for extrusion of left and right needles, N'., N., beyond tip of D. N., right needle withdrawn from D but lying parallel to its original position. N'., anterior part of left needle; the remainder lies concealed within D. N.P., N.P'., piston (?) enlargements on right and left needles. S., stout chitinous piece on dorsal surface of D. The arrows show the direction in which the muscles pull the levers when the sting is employed. The poison sac and duct are not shown.

prepares the way by penetrating the tougher epidermis, and guides the really effective weapons into the softer and more vascular tissues beneath. The whole apparatus consists of three long pieces actuated by a complicated set of muscles and levers, and connected

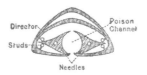

Fig. 19. Diagrammatic cross-section of a wasp's sting in the region of the letter D., in Fig. 18

by a tube with the poison-bag. These three pieces are the "director," and two three-sided "needles," whose opposed sides are concave, and which bear six

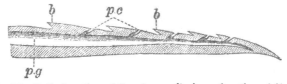

Fig. 20. Optical section of tip of a needle from the sting of *Vespa germanica* ♀. Highly magnified

b., barbs. p.c., poison canals perforating bases of barbs. p.g., poison channel running along inner face of needle, cf. Fig. 19. Actual length of the part figured 0·37 mm., breadth at widest part 0·05 mm.

barb-hooks at their apices. The "director" is sharply-pointed posteriorly and is rounded on its upper surface; while on its lower face is a deep groove in which

VIII] SOME STRUCTURAL FEATURES 111

the "needles" slide to and fro: the edges of the groove are turned in so as to form long slots in which the outer ridges of the needles, right and left, are lodged. The "needles" can thus slide freely forwards and backwards within the "director," but cannot easily be separated from it nor drop out of their respective slots. The whole may be compared to an inverted gutter-pipe, whose sides have been pinched together towards the hinder end, and whose edges have been fitted with a turned-in flange partly enclosing the concavity and lodging two rods within it (*vide* figs. 18, 19).

At its front end the "director" is enlarged, and its groove widens out: in this region the "needles" become free from the inturned flange, but they are still kept in contact and parallel with the margin of the "director" by a double series of peculiar hairs, about fifty in number, situated on its inner face near the lower edge on each side. The outer edge of the "needle" runs between the upper and lower series of hairs, and is thus retained in position. Within the tapering hinder part of the director these hairs are replaced by studs, which cluster in a group in the anterior portion, but are isolated and rather irregularly disposed in the narrower posterior region. These studs prevent any displacement of the "needles" in the passage through the director. Forwards the "needles" diverge from one another and bend rather abruptly upwards: between them is interposed a bulbous mass of muscles (not shown in the figure) attached to the front end of the director. Finally the needles are articulated to strong levers which give attachment to powerful muscles. When the handle of the lever is pulled down by these muscles, in the direction shown by the arrows in the figure, the arm which articulates with the needle is thrust backward and slightly upward, and the needle itself is driven along the groove of the director so as to project beyond it, escaping at a spot just short of its extremity.

In the act of stinging the first incision is made by the point of the "director" itself, which is darted out by the action of its own levers and muscles: a strong piece of horny material (S. fig. 18) on its upper surface at first holds the "director" firmly down, but eventually, when full extrusion has been reached, causes the "director" and "needles" to turn abruptly upwards. This action causes a slight enlargement of the wound beneath the surface. Then by a rapid alternate movement of the levers the "needles" are driven in yet deeper. The two concave inner faces of the "needles" are pressed firmly against each other, partly by the tapering sides of the "director" and partly by the course imparted by the muscles, and so form a closed tube between them. Down this tube the poison is driven by the contractions of the

VIII] SOME STRUCTURAL FEATURES 113

muscular walls of the bag in which the poison is stored. Near its front end each needle bears an enlargement whose surface is covered with numerous fine scales like those of a fish; when, during the protrusion of the "needles," these enlargements reach the narrower part of the "director" they probably act as pistons and sweep the poison along towards the hinder, pointed extremity. It is possible that they also serve to close the triangular cleft between the edges of the director in front of the region where these actually touch each other.

The tube from the poison-bag discharges close against the forward enlargement of the "director" and between the divergent front portions of the "needles," which are thus conveniently placed for receiving the fluid between them. The poison escapes from between the needles at their apices, and also through five minute canals that pass obliquely from the poison groove through the bases of the first five barbs. Neither the "director" nor the "needles" are solid structures; tracheal (breathing) tubes are plainly traceable within their substance, and distinct cavities are visible in transverse sections. It is not probable that the poison produced by all the stinging Hymenoptera is of the same quality and composition: that of some is very deadly in its effect on other insects, whereas in other cases paralysis and not death is caused. Both Carlet and Bordas, who have investigated this matter, state that the poison is formed by the mixture of the secretion of two glands, one of which is acid (formic acid) and the other alkaline. Carlet states that the poison of the Fossors which has merely a stupefactive action is deficient in the alkaline constituent.

The details of structure are, of course, not identical in the stings of all Aculeates: the shape and internal arrangements of the "director," the form of the "needles," and disposition of the levers all vary in different groups: nor are barb-hooks always present, for the "needles" of the Fossors have none. The presence of barbs frequently results in the sting being left in the wound and all its appurtenances being dragged out of the body, with ultimately fatal results to the insect.

The tongue of a bee (*vide* fig. 21, and cf. figs. 9 and 10) forms the central, unpaired structure of an otherwise bilaterally symmetrical and rather complicated set of mouth parts. On the under and posterior side of the head there is a deep recess occupied by the basal part of this apparatus which fits into it. If the whole structure be pushed forward so as to withdraw the basal portion from the headrecess, it will be found that on each side, right and left of the recess, there is articulated a short rod, slightly swollen and flattened at its apex: these rods are really the basal pieces (*cardines*) of the (first)

VIII] SOME STRUCTURAL FEATURES 115

maxillae. Attached to the ends of the cardines are two pieces, the *lora* (*lorum* = rein), which converge towards each other as they pass away from the rods, and eventually meet, so that together they form a V-shaped structure suspended from the rods by the divergent ends of the two limbs of the V. The *lora*

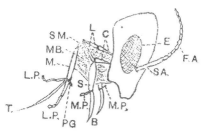

Fig. 21. Diagram of side view of bee's head to show disposition of the mouth parts

B., blades of maxillae = galeae (laciniae of some authors). C., cardines of maxillae. E., eye. F.A., flagellum of antenna. L., lora. L.P., Labial palps. M., mentum. M.B., membranous bag investing bases of mouth parts. M.P., maxillary palps. PG., paraglossa (only one shown). S., stipes of right maxilla. S.A., scape of antenna. SM., submentum. T., tongue.

vary much in length, being much larger in the long-tongued than in the short-tongued bees; in fact in the genera *Halictus* and *Sphecodes* no true *lora* are recognisable. The V formed by the united *lora* is so articulated to the cardines that it can swing right over; thus it can either lie between the cardines with

8—2

its apex pointing towards the thorax, or on the other hand it can be swung forwards so that its point projects some distance beyond them. Now since the tongue and such mouth parts as are directly connected to it are attached to the point where the two *lora* unite, it follows that by this swing of the V they can be thrust forward a distance equal to twice the perpendicular length of the *lora*.

Turning now to the organs articulating with the apex of the V:—next to the united ends of the *lora* is a short piece, the *submentum*, and then a long half tube, the *mentum* (*mentum* = chin), which contains in the trough running down its upper face the soft basal portions of the actual tongue. From the tip of the *mentum* there project (i) in the middle line the tongue itself, technically known as the *ligula*; (ii) to right and left jointed structures, termed *labial palps*: as already stated in an earlier portion of this book, the basal joints of these palps are folded as protecting sheaths round the lower portion of the tongue. The tongue is grooved posteriorly, and its upper surface bears numerous transverse ridges beset with bristly hairs: in the higher bees it has a ladle-like organ, the *bouton* at its tip. At the base of the tongue, and visible only from in front is a pair of little sheaths, concealed by the *mentum* from behind, and known as the *paraglossae*. It may here be remarked parenthetically that the whole of this structure attached to the end of the *lora* corresponds to the *labium*, or united second pair of maxillae, sometimes infelicitously termed the "lower lip" of such an insect as a cockroach with whose anatomy perhaps some of our readers are familiar. There is yet another pair of important organs, viz. the (first) *maxillae*. The basal joints of each of these is the *cardo* already mentioned: to the lower flattened ends of the *cardines* and just external to the articulations of the *lora* there is fastened on each side a more or less rod-shaped piece, the *stipes*. The *stipites* lie to right and left of the *mentum*; and at their far end they carry flattened blades whose inner faces are grooved so that they sheathe, when closed, part of the *mentum* and the base of the tongue. These blades correspond, in my opinion, to the *galeae* of lower insects, though some authorities regard them as the *laciniae*. Springing from the *stipites* at the same level as the blades, but external to them, and projecting outwards, are the jointed *maxillary palps*, right and left: the dimensions of these vary greatly in the different genera of bees. The intervals between the basal parts of the whole of this complex apparatus are bridged over by a membranous material which thus closes the back of the head-recess, and forms round the upper ends of all these mouth parts a kind of bag which leads into the actual mouth.

CHAPTER IX

COLLECTING AND PRESERVING ACULEATE HYMENOPTERA

THE atmospheric conditions necessary for securing specimens of these insects are such as would inevitably tempt most of us to be out in the open rather than indoors. The solitary species are hardly ever to be seen except on bright sunny days; and even when the weather is favourable to their activity, a passing cloud temporarily veiling the sun will cause the majority of them to rest motionless on the ground, and to go within the shelter of their burrow if it be nigh at hand, or to take refuge among the leaves and twigs of neighbouring vegetation. The social forms, impelled perhaps by a sense of duty to the community, continue their work under conditions far less alluring; humble-bees and social-wasps may often be seen on the wing on comparatively dull and cheerless days, or even during light rain. Moreover, these are industrious through a far longer period of the day, and frequently toil for some time after sunset. I have often found wasps on "sugar" placed on trees for the sake of capturing moths when I have visited the baited trunks with a lamp in the early hours of the night; and have witnessed hornets returning to their nest in the dim twilight of a September evening. On

IX] ACULEATE HYMENOPTERA 119

the other hand the solitary species begin work at about 9 or 10 a.m. and begin to "knock off" quite early in the afternoon, so that by 4 or 5 p.m. very few are to be observed.

As a weapon of attack an ordinary butterfly net mounted on a short handle is sufficient; though for taking specimens off flowers, especially of brambles and thorny shrubs, a small net provided with a circular lid capable of closing the mouth of the net-bag is very convenient. The net-ring and its lid of gauze stretched on a second metal ring should be mounted respectively on two metal arms crossed and pivoted together like a pair of large scissors, and the ends of the arms should be provided with the thumb-hole and finger-hole so that the whole apparatus can be used in one hand. It is, however, often possible to pick the specimen off a flower with the finger and thumb when a little experience has been gained in the offensive value of the stings. Having caught "my hare" I usually transfer it to a small glass tube whose cork has been moistened at its inner end with a drop of chloroform before setting out: several specimens may safely be placed in the same tube, but care must be taken that the moisture which evaporates from their bodies does not condense on the sides of the tube, and cause the hairs covering the insects' bodies to become matted together. The best way of preventing this disfigurement is, after a short time, to turn out all the specimens, now stupefied and probably dead, and to wipe out the inside of the tube with the net or handkerchief. The brief exposure of the specimens to a free current of air is usually sufficient to prevent a recurrence of the deposit. Failing chloroform, the ordinary cyanide bottle may be used, provided a wad of blotting-paper has been inserted to prevent moisture from spoiling the specimens. On really hot days no poison is actually necessary, for the heat of the sun itself may be employed to kill the insects: if the tube containing the captured specimens be laid on a flat stone or on a patch of sand fully exposed to the sun's rays the prisoner is very speedily killed. It is extraordinary how very slight a rise of temperature is fatal to these sun-loving creatures. I well remember finding on the sand-dunes of East Norfolk many patches of sloping sand at whose foot lay numbers of dead insects of many kinds: the slopes all had a more or less southerly aspect and were thus exposed to the full heat of the sun. It required but a short period of observation to find out the cause of these piles of dead: after watching for a few minutes I saw a female *Pompilus plumbeus* running about and evidently on the hunt for spiders: there being no danger signal to warn her she chanced to run on to the fatal patch and instantly sprang into the air, only to fall on her back, dead of heat apoplexy. I then placed my hand upon the patch of sand and found its temperature unpleasantly high, but just tolerable. Often too when I have been at a loss for better methods I have killed these insects instantly by placing the tube in hot water of so low a temperature as to be quite comfortable for washing the hands.

In the foregoing chapters the situations in which the various species breed have already been mentioned. Many, of course, when attending to the demands of their own hunger are to be found on the blossoms of many different plants; but it is in the immediate neighbourhood of their nests that all the more dramatic or tragic incidents of their lives are to be witnessed. In the cases of those species which burrow in the ground their presence is usually betrayed either by little heaps of cast up earth, or by the open mouths of the holes themselves; though many conceal these last with such care as to defy detection. A little patience on a suitable day will generally be rewarded by the sight of one or more Fossors or perhaps some species of solitary bee issuing from its burrow, or alighting and making its entry. Should the mouth of the hole be exposed it is probable that the owner is at home: in such a case the tenant may often be tempted up into the daylight again by the aid of a small mirror or a magnifying glass. If the sun happens to be in a favourable position the magnifying glass may be used to focus the rays on the mouth of the hole, and so create an exceptionally fine and warm day—locally; or should the hole be so placed that it is impossible to direct the rays down the hole with the magnifying glass then the mirror may be used to reflect the sunlight into the hole. These devices very seldom fail to entice the insect up into the welcome warmth.

Before being placed in its permanent position in the cabinet or store-box the insect should be "set" in such a way as to display those organs which are of value in identification. But since the insects become rigid soon after death, and since many of them have a knack of dying in very awkward attitudes, it is best to leave them for some days in a relaxing tray before attempting to "set" them. I find one of Newman's patent relaxing boxes most useful for this purpose. When ready for "setting," the insect should be transfixed through the thorax with a very fine pin. It will be found that there is a right place and many wrong places in which to insert the pin; for the muscles which actuate the wings are contained within the hard skeleton of the thorax, and the pin is bound to traverse some of these muscles in its passage. If the right place be found, as it may be by gentle pressure not sufficient to pierce the outside covering, the insertion of the pin will cause the wings to spread out symmetrically and in a position

IX] ACULEATE HYMENOPTERA 123

excellent for subsequent examination. Should the pin be inserted wrongly it is possible sometimes to strap the wings down into good positions with strips of paper, but sometimes no amount of ingenuity or patience will get them to come right. The mouth parts should always be displayed; and here care is necessary as the mandibles are often closed very firmly over the softer maxillae and labium, and in attempting to separate the mandibles with a needle these more delicate structures may easily be injured, and perhaps the entire head torn away from the thorax. The legs should, of course, be spread out from beneath the body; and in the case of male insects the "genital armature" extracted from the end of the abdomen and rendered accessible to the magnifying glass. Great care should be taken to allow the insects to become thoroughly dry before removing them from the "setting-cork"; and it is necessary also to remove from their bodies grains or perhaps masses of pollen before placing them in the cabinet. This last operation is best performed with the aid of a soft paint-brush, first cutting the bristles short so as to render the brush somewhat "stubby." If these precautions are neglected the specimens are very liable to become covered with a growth of filamentous mould, and thus ruined. A further safeguard against mould and animal pests which are so difficult to exclude from a collection of insects is afforded by pinning in the corner of each drawer of the cabinet a small piece of sponge soaked in carbolic acid, and by keeping plenty of naphthaline in each. A wash of dilute formalin painted all over the inner surface of the drawer and allowed to dry before any insects are placed in it is also very effective.

If it is desired to add to the collection the nests of such species as construct nests capable of preservation, it is necessary thoroughly to desiccate the whole structure so as to shrivel up and sterilise all organic matter that may be still within the cells. This process may be accomplished by placing the nest in an open wooden box—metal is too good a conductor of heat and may lead to charring of the parts in contact with it—in a moderately warm kitchen-oven after the culinary operations of the day are completed. In the case of wasps' nests the required temperature is higher than might be expected; for the wrappings of the nest include so much air that the central parts of the nest, i.e. the combs containing the larvae, pupae, and young wasps just ready to emerge, do not receive anything approaching the actual temperature of the oven for a very long time. Should the temperature be insufficient to kill the young wasps, the collector may be disagreeably surprised, as I was on one occasion, on opening the oven door, to find a large number of very lively young wasps crawling over the nest and flying

IX] ACULEATE HYMENOPTERA 125

about in the now nearly cool oven. The moderate temperature merely hastens the emergence of such wasps as are nearly ready to come out from their cocoons. When the nest has been thoroughly baked some formalin should be squirted into it, or some powdered naphthaline shaken well into it before it is put aside in its permanent quarters. Much information regarding the development and life-histories of many species may be obtained by rearing the larvae in captivity. It is not difficult during the summer to mark down spots where Fossors or solitary bees are engaged in excavating their burrows: from such places the larvae or pupae may be dug up with a spade from the depth of about a foot in the autumn, and be taken home for closer observation. Both larvae and pupae have very soft and tender skins and are easily injured: it is thus best to avoid handling them, and to move them only with a soft paint-brush. They should be kept in a box whose bottom is covered with some soft material, so that an accidental fall shall not result in injury. The larvae of many species can be obtained from rotting tree stumps, branches, posts and so forth, and from the stems of brambles. Prunings from wayside hedges will often yield considerable numbers. The chief difficulty in all cases is to preserve the proper degree of moisture in the boxes in which the larvae, etc. are kept through the winter. If they become too dry they shrivel and die, or at any rate are crippled on emergence; on the other hand if kept too wet they are apt to suffer from attacks of mould. It is safest therefore to keep them in a cool place through the winter, and only to expose them to warmth as the season for natural emergence draws near. For this purpose a warm room is better than direct exposure to the sun, from which in their natural surroundings they are, of course, always screened by the surrounding soil or wood fibre, as the case may be. When the insects do emerge they should then be placed where they can enjoy the sunshine and dry the hairs on their bodies, for these are at first generally damp and more or less matted together. Some of these insects live for more than one season in the immature state: hence, if the imagines do not appear in the first summer, it is often worth while to keep the specimens in their boxes through a second winter on the chance of emergences taking place in the summer following.

One of the most satisfactory results of rearing specimens in this manner is the ease with which many parasitic insects and "cuckoos" among the Hymenoptera themselves may with certainty be assigned to their several hosts. To secure this end the nests must, of course, either be kept each in a separate box, or a simpler method may be adopted, namely that of wrapping each bit of bramble stem,

IX] ACULEATE HYMENOPTERA 127

or of rotten wood containing known larvae or pupae in a small piece of gauze or tiffany. Such precautions have the further advantage of making quite sure of nesting sites of species whose larvae have perhaps not been identified at the time of capture; and such are certain to be numerous.

It perhaps "goes without saying" that some degree of patience and of perseverance in the face of disappointments is necessary in these rearing operations; but they are worth the trouble.

Several more or less successful attempts have been made to keep social species under artificial conditions as similar as possible to those that would be found in the open. Wasps have been induced to build their nests in glass-sided cages in principle not unlike the familiar "observatory hives" used by demonstrators in apiculture. The most convenient species for this purpose are those social-wasps which attach their nests to the branches of shrubs or trees. Such nests can be removed whole and without any disturbance of their occupants by carefully cutting through the branch, with as little vibration as possible, in the late evening when all the wasps are at home. A prudent precaution is first to surround the entire nest with the bag of a butterfly net, or some similar material. I have tried repeatedly to get queen wasps to continue their operations in captivity by taking their little nests while the queen was within, and establishing the whole in boxes provided with every convenience for free entry and exit; but the queens appear to resent any transport of their home, and all my attempts with them have ended in failure. But when once the workers have become responsible for the upkeep of the fabric and its occupants transference of the nest does not cause any cessation of their work.

Mr F. W. L. Sladen has recently published an account (*vide* Bibliography) of his work on humble-bees. In his book he explains how by the use of wooden covers for artificial nests these insects may be induced to nest in desired places where they may be studied at leisure.

So far as we are aware no one has ever kept any of the solitary species under the artificial conditions of semi-domestication. The idea of a "home" is probably so feebly developed in the minds of these forms that it is impossible to attach them to any spot save that of their own free choice.

BIBLIOGRAPHY

Wasps, Social and Solitary. By George W. Pechham and Elizabeth G. Pechham. (Constable.)

The Hymenoptera Aculeata of the British Isles. By Edward Saunders. (Reeve.)

Wild Bees, Wasps and Ants. By Edward Saunders. (Routledge.)

Insect Life: Souvenirs of a Naturalist. By J. H. Fabre (Translation). (Macmillan.)

The Life and Love of the Insect, chaps. xv and xvi. By J. H. Fabre (Translation). (Black.)

The Cambridge Natural History, vols. v and vi. By David Sharp. (Macmillan.)

Bees and Bee Keeping. 2 vols. By F. R. Cheshire. (Upcott Gill.)

The Honey-Bee. By T. W. Cowan. (Houlston.)

British Bees. By W. E. Schuckard. (Reeve.) 1866.

Ants, Bees and Wasps. By Sir John Lubbock (Lord Avebury). (Kegan, Paul.)

Mémoires pour servir à l'histoire des Insectes. By Réaumur. 1740.

Nouvelles Observations sur les Abeilles. By François Huber. 1794.

The Life of the Bee. By Maurice Maeterlinck. (Allen.)

The Humble-Bee, its Life-history and how to domesticate it. By F. W. L. Sladen. 1912. (Macmillan.)

INDEX

(*Numbers in italics indicate an illustration on the page*)

Ammophila 31
  campestris 24, 31
  sabulosa 20, *21*-23, 31
Andrena 4
  mouth parts of *59*
  argentata 64
  fulva 64–66
Andrenidae 63–73
Antennae 3
Anthidium 57, 73, 74
  manicatum 85, 86
Anthophila 4, 57–108
Anthophora pilipes 74–76, 87
Ants 3
Apidae 73–108
  mouth parts of 74
Apis 73, 74
  mellifica 95–108
  leg of *105*

Bee, definition of 3, 4
Beetles 2
Bombus 57, 73, 74, 87–95
  mouth parts of *61*
  agrorum 88
  lapidarius 88, 93
  muscorum 93
  terrestris 88
  venustus 88
Bumble-bees 73

Cardines *59*, *115*–117
'Cells,' number of in wing 2
Chelostoma 74, 78

Chrysididae 40
Chrysomyia polita 35
Coelioxys 84
Collecting 118–128
Colletes, mouth parts of *59*, 60–62, 73
  succincta 62
Corbiculum *105*, 106
Crabro 28, *29*–36
  cetratus 32
  clavipes 35
  clypeatus 33
  cribrarius 33, 35
  dimidiatus 35
  gonager 32
  interruptus 33
  leucostomus 35
  palmarius 32
  peltarius 33
  quadrimaculatus 35
  scutellatus 33
  tibialis 32
  wesmaeli 35
Cuckoo' bees 58, 62, 64, 66, 70, 71, 73, 75, 76, 78, 84, 93, 94, 115

Dasypoda hirtipes *71*–73
Diploptera 5, 36–57
Dragon-flies 2
Drone 107

Earwigs 2
Entomognathus 31, 32

INDEX

Epeolus 62, 73
Eucera 73
  longicornis 86, 87
Eumenes coarctata 36

Formic acid 103, 114
Fossores 5–36

Galeae *59, 115,* 117
Gall-flies 2
Grass-hoppers 2

Hairs 3, 58, *105,* 106
Halictus 64, *67,* 70, 115
  lineolatus 68
  morio 69
  sexcinctus 68
Hind leg 3, *105*
Honey 102–104
Honey-bee 73, 95–108
Hornet 40, 41
Humble-bees 73, 87–95
  economic importance of, 94, 95

Ichneumon-flies 2

Jewel-wasps 40

Labium 116, 117
Laciniae *59, 115,* 117
Leaf-cutter bees 73, *79, 80–85*
Lepidoptera 2
Ligula 116
Long-horned bee 73
Lora *115,* 116
Lycosa picta 12

Mandibles 3, 123
Mason bees 73
Maxillae 3, *115,* 116, 117, 123
Mayflies 2
Megachile 73, 74, 78, *79, 80–85*

Megachile circumcincta 81
  maritima 83
Melecta armata 75, 76
Mentum *59, 115*–117
Metatarsus 3, *105*
Mouth parts *59, 61,* 114, *115,* 123

Nectar 102
Nomada 58, 64, 66

Odynerus 36–40
  parietum 37–39, 40
  spinipes 37
Osmia 73, 74, *76–78*
  bicolor 77
  fulviventris 78
  leucomelana 77
  rufa 76, 77
Oxybelus quadrinotatus 27
  uniglumis 24, *25–28,* 31

Paraglossae *59, 115,* 116
Pegomyia inanis 54
Philodromus fallax 13
Poison bag 1, 112, 113
Pompilids 5–19
  sites of nests 6
  prey of 6
Pompilus plumbeus 8–14, 120
  rufipes 14, 18, 19
  viaticus 14–17
Preserving 118–128
Propolis 100
Prosopis 60, 62, 63
Psithyrus rupestris 93, 94

Rhipiphorus paradoxus 55
Rima of *Halictus* 67
Royal jelly 100
Ruby-wasps 40

Sarapoda bimaculata 86, 87

Saw-flies 2
Social-wasps 40–57
  nest of *43–46*–57
Sphecodes 58, 64, 70, 71, 115
Sphegidae 19–36
Stelis 78
Sting 1, 2, 108, *109*, *110*–114
Stipes *115*, 117
Stylops 69, 70
Submentum *115*–117
Swarms 97–99

Vespa austriaca 42
  crabro 41
  germanica 41
    sting of *109*, *110*–114
Vespa norwegica 42
  rufa 41, 42
  sylvestris 42
  vulgaris 41
Vespidae 40–57
Volucella inanis 54

Wasp, definition of 3, 4
Wasp's nest, temperature within 51
Wasps, uses of 56, 57
Wax 57, 89, 96, 104
Wings 2
  nervures and cells of 28, *29*–31
Wool-carder bee 73

CPSIA information can be obtained at www.ICGtesting.com
Printed in the USA
BVOW012020140312

285120BV00001B/31/P